Nanocharacterization
Techniques

Nanocharacterization Techniques

Edited by

Alessandra L. Da Róz

Marystela Ferreira

Fabio de Lima Leite

Osvaldo N. Oliveira Jr.

William Andrew is an imprint of Elsevier
The Boulevard, Langford Lane, Kidlington, Oxford, OX5 1GB, United Kingdom
50 Hampshire Street, 5th Floor, Cambridge, MA 02139, United States

Library of Congress Cataloging-in-Publication Data
A catalog record for this book is available from the Library of Congress

British Library Cataloguing-in-Publication Data
A catalogue record for this book is available from the British Library

ISBN: 978-0-323-49778-7

For information on all William Andrew publications visit our website at
https://www.elsevier.com/books-and-journals

Working together
to grow libraries in
developing countries

www.elsevier.com • www.bookaid.org

Publisher: Matthew Deans
Acquisition Editor: Simon Holt
Editorial Project Manager: Charlotte Kent
Production Project Manager: Lisa Jones
Designer: Greg Harris

Typeset by Thomson Digital

Table of Contents

List of Contributors

Priscila Alessio

State University of São Paulo, São Paulo, Brazil

Alvaro E. Aliaga

State University of São Paulo, São Paulo, Brazil

Pedro H.B. Aoki

State University of São Paulo, São Paulo, Brazil

Aldo F. Craievich

Institute of Physics, University of São Paulo, São Paulo - Sp, Brazil

Marcelo de Assumpção Pereira-da-Silva

Institute of Physics of São Carlos, IFSC-USP; Central Paulista University Center, UNICEP, São Carlos, SP, Brazil

Fábio de Lima Leite

Federal University of São Carlos, Sorocaba, Brazil

Mario de Oliveira Neto

Institute of Biosciences, University of São Paulo State, Botucatu - Sp, Brazil

Daiana K. Deda

Federal University of São Carlos, Sorocaba, Brazil

Marystela Ferreira

Federal University of São Carlos (Ufscar), Sorocaba, Sp, Brazil

Fabio A. Ferri

Institute of Physics of São Carlos, IFSC-USP, São Carlos, SP; Federal University of Lavras, Lavras, MG; Federal University of São Carlos, São Carlos, SP, Brazil

Leonardo N. Furini

State University of São Paulo, São Paulo, Brazil

Pâmela S. Garcia

Federal University of São Carlos, Sorocaba, Brazil

Guinther Kellermann

Department of Physics, Federal University of Paraná, Curitiba - Pr, Brazil

Diego G. Lamas

Conicet/School of Science and Technology, National University of San Martín, San Martín, Argentina

Carlos J. Leopoldo Constantino

State University of São Paulo, São Paulo, Brazil

Celina M. Miyazaki

Federal University of São Carlos (Ufscar), Sorocaba, Sp, Brazil

Ana P. Ramos

University of São Paulo, Ribeirão Preto, São Paulo, Brazil

Flávio M. Shimizu

São Carlos Institute of Physics, University of São Paulo, São Carlos, Sp, Brazil

Ronald Tararam

Multidisciplinary Center for the Development of Ceramic Materials, São Paulo State University, Araraquara, Brazil

José A. Varela

Multidisciplinary Center for the Development of Ceramic Materials, São Paulo State University, Araraquara, Brazil

Editor Biographies

Alessandra L. Da Róz completed a degree in Sciences, with a major in Chemistry, and obtained her PhD in Materials Science and Engineering at the University of São Paulo, in São Carlos, Brazil. After a postdoc and a researcher positions at the São Carlos Institute of Physics, University of São Paulo, and at the Federal University of São Carlos, she now works in the coordination for research and development at the Federal Institute of Education, Science and Technology of São Paulo (IFSP) in Itapetininga, Brazil. Her main areas of research include polymers and their applications, such as in the processing of natural polymers, lignocellulosic biomass, and solid fuels.

Marystela Ferreira obtained her BSc in Chemistry in 1993, and MSc and PhD in Physical Chemistry in 1996 and 2000, respectively, at the University of São Paulo, São Carlos, Brazil. Her main fields of interest are preparation and characterization of nanostructured thin films for sensing, and layer-by-layer and Langmuir–Blodgett films of polymers for sensing different analytes in environmental and biological samples. She is a lecturer at the Universidade Federal de São Carlos, Sorocaba, Brazil, since 2007.

Fabio de Lima Leite obtained his PhD in Materials Science and Engineering from University of São Paulo, Brazil. Between 2006 and 2009, he was a Postdoctoral Fellow at the São Carlos Institute of Physics (IFSC-USP) in collaboration with Embrapa Agricultural Instrumentation. He was a FAPESP Young Researcher Fellow (2009–12). He has collaborated with Prof. Dr. Alan Graham MacDiarmid, winner of the Nobel Prize in Chemistry 2000, with whom he published the first article in the *Journal of Nanoscience and Nanotechnology*, in 2009. He is currently Adjunct Professor at the Federal University of São Carlos (UFSCar), Campus Sorocaba, and Coordinator of the Research Group in Nanoneurobiophysics and Future Scientist Program. He has conducted research in the areas of nanoscience and nanotechnology, with emphasis on nanoneuroscience and medical nanobiophysics. Currently, he leads research in neurological disorders, which has launched a new research area entitled "nanoneurobiophysics."

Osvaldo N. Oliveira Jr. completed his PhD at the University of Wales, Bangor, United Kingdom. He is a professor at the São Carlos Institute of Physics, University of São Paulo. His main areas of expertise are in nanostructured organic films, and natural language processing. He is a member of the São Paulo State Academy of Sciences. He is a member of the editorial board of three journals, and is also associated editor of the *Journal of Nanoscience and Nanotechnology*. He received the Scopus Award 2006 awarded by Elsevier in Brazil and Capes, as 1 of 16 outstanding Brazilian researchers, based on the number of publications and citations.

1

Scanning Electron Microscopy

Marcelo de Assumpção Pereira-da-Silva*,**, Fabio A. Ferri*,†,‡

*INSTITUTE OF PHYSICS OF SÃO CARLOS, IFSC-USP, SÃO CARLOS, SP, BRAZIL;
**CENTRAL PAULISTA UNIVERSITY CENTER, UNICEP, SÃO CARLOS, SP, BRAZIL;
†FEDERAL UNIVERSITY OF LAVRAS, LAVRAS, MG, BRAZIL;
‡FEDERAL UNIVERSITY OF SÃO CARLOS, SÃO CARLOS, SP, BRAZIL

CHAPTER OUTLINE

1 Introduction

Microscopy is a technique used to visualize structures that cannot be observed with the naked eye. Its primary purpose is to form an image of the area intended to be observed.

Microscopy techniques allow visualization of structures present within the sample or on its surface, depending on the technique used and the characteristics of the sample. To

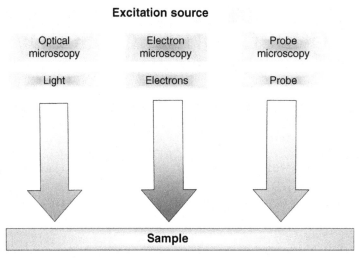

FIGURE 1.1 Types of commonly used excitation sources in microscopy techniques.

visualize a sample, techniques to improve the resolution capacity of the human eye are used, which is on the order of 0.2 mm.

Until the first quarter of the 20th century, samples were observed using visible light with so-called optical microscopes, and this technique depended on the development of lens production techniques with appropriate optical qualities to achieve a resolution limit of half of the shortest wavelength of visible light.

Among the various ways to classify microscopy techniques, one of them relates to the excitation source for the sample. Currently, the three most used microscopy techniques—light microscopy, electron microscopy, and probe microscopy—use light, electrons, and a probe as excitation sources, respectively (Fig. 1.1).

When the excitation source hits the sample, various types of interactions take place, resulting in the emission of different signals. In some cases, these signals are emitted from the same side on which the sample's excitation source is incident and are called scattered signals. In other cases, the signals are emitted from the opposite side to that on which the sample's excitation source is incident and are referred to as transmitted signals. With an electron excitation source, the technique dedicated to capturing transmitted signals is called transmission electron microscopy, and the technique that captures scattered signals is called scanning electron microscopy, the subject of this chapter (Fig. 1.2).

2 Scanning Electron Microscope

The scanning electron microscope (SEM) consists of two major parts, the column and the cabinet (Fig. 1.3). The column is the extension that the electrons traverse from their emission until they reach the sample, where the installed detectors will capture the scattered

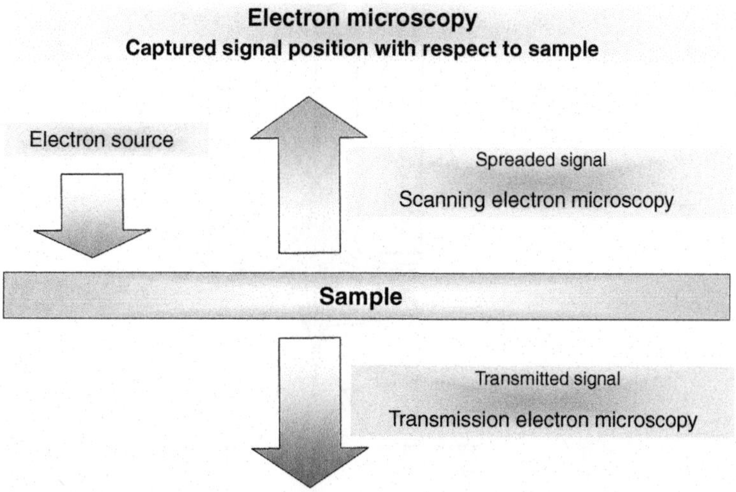

Electron microscopy
Captured signal position with respect to sample

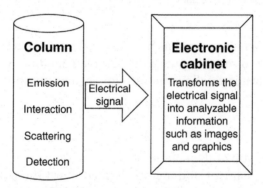

FIGURE 1.2 The position of the captured signal defines the two types of electron microscopy.

FIGURE 1.3 The electron column shows all of the elements that pertain to the signals from their emission until their capture. In the cabinet, the signals are processed for easy display.

signals resulting from the interaction between the electrons and the sample. The detectors are energy transducers that transform one type of signal into an electrical signal, which is sent to the control cabinet. The control cabinet has electronic systems able to quantify the electrical signals sent by the detectors and turn them into analyzable information such as images and graphs.

2.1 Vacuum

In an SEM, a vacuum is required in the electron column and sample chamber because electrons can travel only a small distance through air. The vacuum is produced through a turbomolecular pump backed by a mechanical rotary pump. The turbomolecular pump

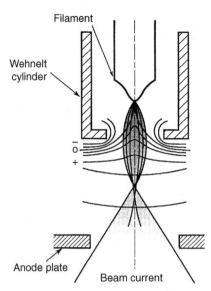

FIGURE 1.4 Schematics of the electron source or electron gun. *Adapted from Goldstein J, Newbury DE, Joy DC, Lyman CE, Echlin P, Lifshin E, Sawyer L, Michael JR: Scanning electron microscopy and X-ray microanalysis, New York, 2007, Springer.*

starts operating only after a vacuum has been created by the mechanical pump, which is used to preevacuate or to roughly pump the sample chamber. After the preestablished vacuum, a valve is activated to allow the turbomolecular pump to evacuate the sample chamber.

2.2 Electron Gun

The electron gun at the top of the column is the electron source (Fig. 1.4). Electrons are emitted from a heated filament and are accelerated down the column. There are three electrically isolated parts in the gun (Goldstein et al., 2007). (1) The first part consists of a filament that emits the electrons (cathode) and creates a cloud of electrons around itself. (2) The second part consists of a metal cylinder (Wehnelt) with an aperture involving the emitter. This cylinder controls the number of electrons that leave the gun. A negative potential is applied to this cylinder, and around its aperture, field lines are formed, which will reduce the diameter of the electron cloud along the gun. (3) The third part comprises a disc with an aperture (anode) that accelerates the electrons at voltages of 0.5 and 30 kV. The disk with an aperture is placed to form an electric field with the cylinder capable of accelerating electrons along the gun.

2.2.1 Filament Types
The following are two types of electron guns: guns with a thermionic emission filament, having a filament of tungsten or lanthanum hexaboride (LaB_6), and guns with field emission filaments, both thermal and cold (Fig. 1.5).

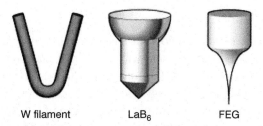

W filament   LaB$_6$ FEG

FIGURE 1.5 Illustration of the main types of filaments. *Adapted from http://lli155-94.members.linode.com/myscope/ sem/practice/principles/gun.php; Goldstein J, Newbury DE, Joy DC, Lyman CE, Echlin P, Lifshin E, Sawyer L, Michael JR: Scanning electron microscopy and X-ray microanalysis, New York, 2007, Springer.*

The most commonly used filament is the tungsten filament, which is heated to a temperature of 2800 K during operation. The high temperature provides kinetic energy for the electrons to overcome the surface energy barrier and leave the filament (Table 1.1).

The lanthanum hexaboride (LaB$_6$) filament requires less energy for the electrons to escape from the filament since it has a surface work function of 2 eV while tungsten has a value of 4.5 eV. The LaB$_6$ filament provides greater beam intensity, but it must operate at a higher vacuum level. In the filaments for field emission, electrons are pulled from the filament surface by the tunneling effect instead of the thermionic effect through the application of a very high electric field that allows the electrons of the filament to overcome the surface energy barrier.

The field emission filaments made of a tungsten crystal with a very fine tip, on the order of 100 nm, provide an electron intensity 10,000 times greater than that of the common tungsten filament and at least 100 times higher than that of the LaB$_6$ filament. The thermal field emission filaments operate in a temperature range of 1600–1800 K and provide an emission current with low noise. The cold field emission filament operates at room temperature, has a very small power distribution (0.3–0.5 eV), and is very sensitive to residual ions that collide with the filament, causing emission instability. This filament operates at a vacuum of 10^{-10} Torr and requires frequent maintenance to remove residues deposited by ions from the surface of the filament (Leng, 2008).

Table 1.1   Comparison of the Electron Sources

	Filament		Field Emission	
	Tungsten	**LaB$_6$**	**Thermal**	**Cold**
Operating temperature (K)	2800	1900	1800	300
Brightness [A/(cm sr kV)]	10^4	10^5	10^7	2×10^7
Required vacuum (Torr)	10^{-4}	10^{-6}	10^{-9}	10^{-10}
Energy distribution (eV)	2.5	1.5	1.0	0.25
Lifetime (h)	100	1000	5000	2000
Filament regeneration	No	No	No	Every 6–8 h
Emission current/area (A/cm^2)	3	30	5.300	17.000

Adapted from www.tedpella.com/apertures-and-filaments_html/yps-schottky.htm.

2.3 Electron Column

The electron column is located just below the disc aperture. In the electron column, there are the condenser lenses, objective lenses, and scanning lenses. The lenses nearest to the electron gun are called condenser lenses, while lenses nearest to the sample are called objective lenses. The condenser lenses, magnetic lenses located below the electron gun, are used to reduce the electron beam to a small cross-section of 5–50 nm in diameter from an initial cross-sectional diameter more than 1000 times greater. The electron beam enters as a cylindrical shape with a diameter on the order of millimeters and is condensed to form a cone whose vertex is a few nanometers. Next, the objective lenses alter the vertical position of the vertex and allow focusing through the different vertical positions of the sample. The function of the objective lenses is to move the smallest cross-section of the beam up and down to find the sample surface, which corresponds to focusing the image. The scanning lenses deflect the electron beam in both directions over the sample surface, causing the electron beam to hit and interact with an array of sample points.

The final aperture is a platinum disc with a small hole (±100 μm diameter) located just before the sample chamber, and its function is to limit the angular width (solid angle) of the electron beam to reduce the effects of spherical aberration and improve the depth of field in the image.

Another lens system is responsible for beam scanning, and its scanning coils are used to deflect the beam across the sample in sync with the video monitor that displays the image.

2.4 Sample Chamber

The sample chamber connected to the vacuum line is located just below the objective lenses. Furthermore, the moving sample stage, electron signal detectors, and X-ray detectors are located within the chamber.

To insert a sample into the chamber, the electron beam must be turned off; the vacuum is released from the sample chamber and dry nitrogen is vented into it. Then, a new sample can be inserted into the sample stage, and the chamber is again evacuated. A good vacuum is very important, and every place in the evacuated area, including the sample, should be handled only with dust-free gloves.

The stage and detectors are located inside the chamber. The stage is where the samples are placed for analysis. The detectors are responsible for capturing the signals that were scattered by the sample and act as transducers of these signals into an electrical signal.

The electrical signal will be sent to the control cabinet that has electronic systems able to quantify the electrical signals sent by the detectors and turn them into analyzable information such as images and graphs.

3 Using the SEM

The SEM is used to observe and modify the sample's surface. It is used to capture and interpret some signals emitted during the interaction of the electron beam with the sample (Fig. 1.6). Among these signals, there are electrons [Auger electrons, secondary electrons

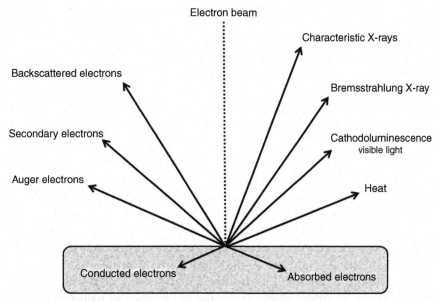

FIGURE 1.6 Main signals emitted as a result of the interaction between the electron beam and the sample. *Adapted from Goldstein J, Newbury DE, Joy DC, Lyman CE, Echlin P, Lifshin E, Sawyer L, Michael JR:* Scanning electron microscopy and X-ray microanalysis, *New York, 2007, Springer.*

(SEs), and backscattered electrons], X-rays (characteristic X-rays and Bremsstrahlung X-ray radiation), light (ultraviolet, visible, and infrared), heat, electrons conducted through the sample, and electrons absorbed by the sample. With some of these signals, it is possible to observe and characterize the sample in terms of its (1) surface morphology, (2) structural organization, and (3) chemical composition.

Surface modification occurs because the electron beam, when it has sufficient energy, is able to locally change the sample's surface material, which may generate nanometer-sized structures. The most commonly used technique for sample surface modification using an electron beam as a writing and design tool with a resolution of a few nanometers, known as electron beam lithography (EBL).

4 Developments in Scanning Electron Microscopy

The developments in scanning electron microscopy have brought important changes to the observation of surface morphology, including the use of low-voltage scanning electron microscopes (LVSEMs), the use of variable pressure or environmental scanning electron microscopes (ESEMs), the use of sources capable of providing greater brightness than scanning electron microscopy with field emission filaments [field emission scanning electron microscopes (FESEMs)], and electron detectors within the lenses.

4.1 Electric Charge Buildup in the Sample

In general, the samples examined in an SEM must be electrically conductive to minimize charge buildup on the sample caused by the electron beam. The charge buildup can degrade the sample and distort the image data (Sawyer et al., 2008). During imaging, electrons are continuously bombarding the sample, and a negative charge can build up in areas of the sample under the beam. This negative charge, when sufficiently large, can deflect the incident and emitted electrons, thus ruining the image. To prevent this effect, each sample must be electrically conductive so that the current deposited by the electron beam on the sample can pass through the sample stage to the electrical ground. Some samples, such as metals, are already conductors; however, other samples, such as ceramics, polymers, and biological materials, are not conductive. Thus, the sample surface is coated with a thin layer of an inert conductive substance, such as gold or carbon, using evaporation or plasma coating devices. Moreover, to ensure the electrical contact between this conductive layer and the metallic holder of the sample, a conductive silver or carbon ink is used, ensuring good grounding of the sample. In addition to the conductive covering, other approaches can be employed to minimize the buildup. One of these is to increase the chamber's pressure, thereby balancing the electrical buildup with the gas molecules using an ESEM. Another approach is to reduce the voltage of the electron beam using an LVSEM.

4.2 Nanoassembly

The charge buildup can also be used as an auxiliary technique in the assembly of nanoparticles on an insulating substrate (Fig. 1.7A and B). The surface is electrically charged with the beam finely focused on a predetermined pattern; afterward, the nanoparticles are deposited and will attach to the substrate through electrostatic attraction only to those regions charged by the electron beam. The process can be subdivided into the following steps: (1) creation of a charge pattern on an insulating substrate using a finely focused electron beam, (2) deposition of charged nanoparticles from the gaseous phase of an ESEM on the pattern, and (3) inspection of the deposit using an ESEM. Fig. 1.7A shows the schematics of the process, while Fig. 1.7B shows the result of the deposition of palladium nanoparticles on an Si_3N_4 substrate (Zonnevylle et al., 2009).

4.3 Electron Detectors

Electron detectors provide two types of contrast, topographic contrast and compositional contrast (atomic number contrast). The detector of SEs, which forms the image of SEs, known as the Everhart–Thornley detector, is the most common type of electron signal n and provides topographic contrast. The detector is sensitive to backscattered electrons and SEs. The backscattered electrons are sufficiently energetic to trigger the detector. The low-energy SEs are driven in the direction of the detector by the voltage potential of a positively charged wire grid collector. The electrons are accelerated until they reach the detector/scintillator, producing light. This light, after amplification in a photomultiplier

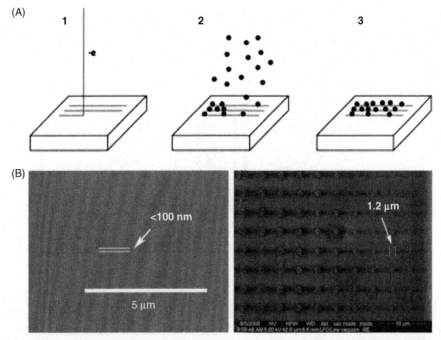

FIGURE 1.7 (A) Nanoassembly process: creation of a charge pattern and deposition of charged nanoparticles; (B) nanoassembly result: palladium nanoparticles on an Si_3N_4 substrate. *Adapted from Zonnevylle AC, Hagen CW, Kruit P, Schmidt-Ott A: Directed assembly of nano-particles with the help of charge patterns created with scanning electron microscope,* Microelectron Eng *86(4–6):803–805, 2009.*

tube, produces an electrical signal, which will modulate the brightness on the monitor displaying the image.

The backscattered electron detector performs the backscattered electron imaging (BSEI), which provides the atomic number contrast (Fig. 1.8). Regions with a high average atomic number appear brighter than regions with a lower average atomic number. Thus, regions with higher average atomic numbers will provide a higher brightness, while regions separated by elements with a difference of only one atomic number unit will produce a low contrast. Combining BSEI with the technique of energy-dispersive X-ray

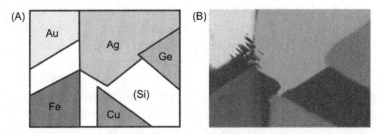

FIGURE 1.8 Backscattered electron image showing the difference in contrast due to the atomic number: (A) observed material and (B) image of backscattered electrons. *Adapted from Kim H, Negishi T, Kudo M, Takei H, Yasuda K: Quantitative backscattered electron imaging of field emission scanning electron microscopy for discrimination of nano-scale elements with nm-order spatial resolution,* J Electron Microsc *59(5):379–385, 2010.*

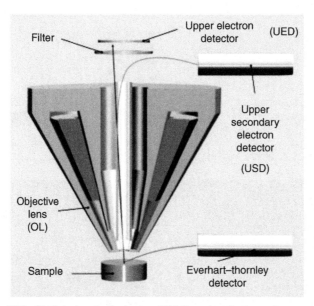

FIGURE 1.9 **Setup of an SEM with detectors in the lenses.** *SEM*, Scanning electron microscope. *Adapted from Asahina S, Uno S, Suga M, Stevens SM, Klingstedt M, Okano Y, Kudo M, Schueth F, Anderson MW, Adschiri T, Terasaki O: A new HRSEM approach to observe fine structures of novel nanostructured materials,* Microporous Mesoporous Mater *146(1–3):11–17, 2011.*

spectroscopy (EDS), with voltages varying between 1 and 10 keV in an FESEM, it is possible to analyze thin films of nanometer thicknesses. The BSEI assists in the lateral resolution to identify nanometer-sized particles through atomic number contrast (Kim et al., 2010).

To map the actual surface of the sample, if possible, only the SEs generated by the incident beam (SE1) should be detected since only these are directly generated near the beam's impact point. These electrons can be very efficiently detected using sensors in the lenses (Griffin, 2011), whose detection results from the geometric positioning of the electrons within the beam's path and from the combination of electrostatic and electromagnetic lenses mounted next to the objective lenses. Fig. 1.9 shows the setup for these types of detectors.

The detector is placed above the objective lenses in the beam's path. Some settings use a filter to attract or repel electrons inside and around the lenses, which allows selection of topographical or compositional contrast. Electrostatic lenses (EL) placed inside the objective lenses are used to slow down the primary beam and choose the beam's optimal impact energy on the sample. The SEs generated at the point of impact on the sample are reaccelerated to higher energies by the EL and refocused by electromagnetic lenses, both placed inside the objective lenses. Another employed technique consists of controlling the voltage applied to the sample to slow down the original energy from the beam and reduce the energy on impact in an attempt to reduce the charge on the sample (Asahina et al., 2011).

A nanopore network is visible on a mesoporous silica structure (Fig. 1.10A and B) (Stevens et al., 2009). Pores with details on the order of a few nanometers are clearly

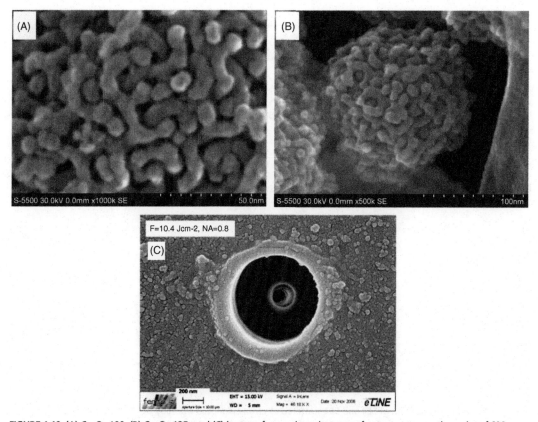

FIGURE 1.10 (A) Co_3O_4-100, (B) Co_3O_4-135, and (C) image of secondary electrons of a nanocrater on the order of 600 nm. *Parts A and B adapted from Tuysuz H, Lehmann CW, Bongard H, Tesche B, Schmidt R, Schuth F: Direct imaging of surface topology and pore system of ordered mesoporous silica (MCM-41, SBA-15, and KIT-6) and nanocast metal oxides by high resolution scanning electron microscopy,* J Am Chem Soc *130(34):11510–11517, 2008. Part C adapted from Delobelle B, Courvoisier F, Delobelle P: Morphology study of femtosecond laser nano-structured borosilicate glass using atomic force microscopy and scanning electron microscopy,* Opt Lasers Eng *48(5):616–625, 2010.*

visualized (Fig. 1.10C), and the image from the SEs also provides depth perception since in these regions, due to the capturing difficulty, they have lower intensities (Delobelle et al., 2010).

5 Low-Voltage Scanning Electron Microscopy

When we perform high-resolution microscopy (Liu, 2000) with a high-energy beam, it is necessary to distinguish between SEs generated by the incident beam (SE1) and SEs produced by the electrons scattered by the sample (SE2) (Cazaux, 2004) (Fig. 1.11) since these two components have different spatial distributions (Joy, 1991; Joy and Pawley, 1992).

The SE1 signal is generated in a region a few nanometers from the beam's impact point, but the SE2 signal comes from a distant region a few hundred nanometers away from the

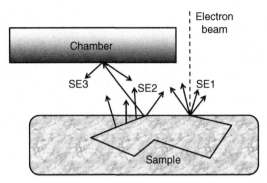

FIGURE 1.11 Source of signals SE1, SE2, and SE3 (Goldstein et al., 2007; Joy, 1991).

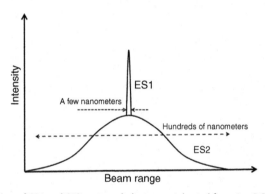

FIGURE 1.12 Spatial distribution of SE1 and SE2 scattered electrons. *Adapted from Joy DC: The theory and practice of high-resolution scanning electron microscopy,* Ultramicroscope 37(1–4):216–233, 1991.

electron beam's impact point (Fig. 1.12) (Joy and Joy, 1996). The intensity of the SE2 signal is higher than the intensity of the SE1 signal; thus, the high-resolution information supplied by SE1 electrons is only a small fraction of the information provided by the low-resolution SE2 electrons, and the signal to noise ratio of the high-resolution SE1 electrons is poor (Joy, 1991).

When we use LVSEM in the voltage range from 0.5 to 5 keV, it is possible to obtain an appropriate combination of the number of incident electrons (nie) and the sum of the number of SEs (nse) plus the number of backscattered electrons (nbe). Considering the yield of emitted electrons as the sum ($\sigma = \eta + \delta$) of the yield of SEs (η = nse/nie) and backscattered electrons (δ = nbc/nie) within a critical energy range (Cer) of the beam's electrons, between Cer1 and Cer2 it is possible to obtain coefficients $\sigma > 1$, avoiding electrical buildup on the sample since the number of emitted electrons is greater than the nie (Fig. 1.13).

When $\sigma > 1$, the sample's surface tends to become positively charged, creating a stable situation in which the sample begins to attract the beam's incident electrons. If energies exceeding Cer2 or less than Cer1 are used, electrons will be accumulated on the sample, which may, in some cases, destroy the sample by means of cracks or change the sample

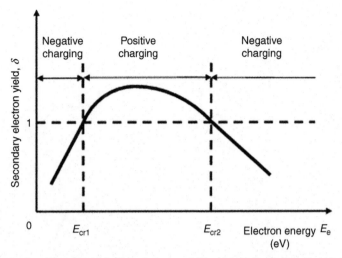

FIGURE 1.13 Electron yield as a function of the beam's energy. *Adapted from Komoda H, Yoshida M, Yamamoto Y, Iwasaki K, Nakatani I, Watanabe H, Yasutake K: Novel charge neutralization techniques applicable to wide current range of FIB processing in FIB-SEM combined system,* Microelectron Reliability *46(12):2085–2095, 2006.*

into a mirror since the incident electrons will be repelled by the negative charge on the sample (Goldstein et al., 2007; Komoda et al., 2006; Seiler, 1983). Since the images are a combination of the beam's diameter, beam's current, extent of the beam's interaction with the sample, and yield of the detection system, at small voltages, there is the advantage of obtaining images in many electrically insulating samples without the need to apply a conductive coating (Boyes, 1998; Erlandsen et al., 2003).

Considering the earlier information, a small beam diameter, as performance criteria for obtaining high-resolution images, although valid, should be combined with lower voltages to maximize the contrast in images of SEs (Schatten, 2011). This low-voltage technique was not available in the equipment a few years ago. LVSEMs became possible due to the development in SEM instrumentation that allowed easy operation using low-voltage beam energies less than 5 keV. The modern scanning microscopes can operate with low voltages between 0.2 and 5 keV since imaging in this energy range was made possible (Muray, 2011).

Fig. 1.14 shows a LVSEM image of the indentation marks in irradiated and nonirradiated areas, where fissure details on the order of 20 nm can be observed (Oono et al., 2011).

6 Environmental Scanning Electron Microscopy

The ESEM is defined as an SEM that can operate with a chamber containing samples and gas, in addition to a vacuum, with a pressure level that can hold wet samples (609 Pa), liquid water, and living samples (Ahmad et al., 2009, 2012; Kaminskyj and Dahms, 2008).

In addition, the gas pressure offers the advantage of acting as a means of charge dissipation, and with the correct choice of the gas composition, it may be used as a means

(A)   (B)

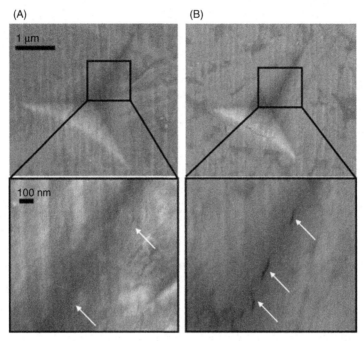

FIGURE 1.14 LVSEM images of indentation marks in (A) nonirradiated area and (B) irradiated area showing fissures on the order of 20 nm. *LVSEM*, Low-voltage scanning electron microscope. *Adapted from Oono N, Kasada R, Higuchi T, Sakamoto K, Nakatsuka M, Hasegawa A, Kondo S, Matsui H, Kimura A: Irradiation hardening and microstructure evolution of ion-irradiated Zr-hydride,* J Nucl Mater *419(1–3):366–370, 2011.*

of detection as well as for microfabrication, which expands the possibilities for SEM that were limited by the use of a vacuum (Bogner et al., 2007).

The ESEM technology uses pressure-limiting apertures (PLAs) (Fig. 1.15) to restrict gas flow between the sample chamber region, with pressures up to 1 atm, and the evacuated region of the electron gun as well as the electron column. Among the PLA, a differential vacuum pumping system is used allowing various pressure differences of several orders of magnitude to be created, making it possible for the chamber pressure to be higher than the pressure in the gun. Successive stages can be added to obtain a high vacuum in the electron gun [pressure at the second stage of the column (electron gun) of the ESEM (P3)] allowing the use of LaB_6 guns or of field emission. Gas refueling [pressure at the gas entrance of the ESEM (P0)] is performed by an external gas inlet control valve (CV), and since the flow of gas through the PLA1 is very low, the velocity of the gas in the sample chamber is also very low; therefore, there are no large pressure gradients within the sample chamber (P1 pressure region).

There is a significant loss of electrons, which begins in the region between PLA1 and PLA2 [pressure at the first stage of the column of the ESEM (P2)]; however, the greatest loss of electrons occurs in the sample chamber (P1 pressure). The electron beam should reach the sample with minimum loss. The electron scattering process through the gas follows a

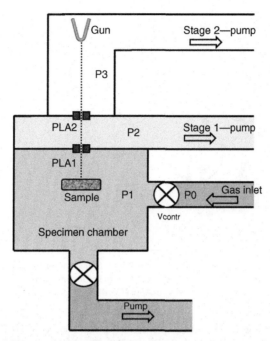

FIGURE 1.15 An ESEM of variable or ambient pressure. *ESEM,* Environmental scanning electron microscope; *PLA,* pressure-limiting aperture. *Adapted from Danilatos GD: Design and construction of an atmospheric or environmental SEM. 1,* Scanning *4(1):9–20, 1981; Danilatos GD: Design and construction of an atmospheric or environmental SEM. 3,* Scanning *7(1):26–42, 1985; Danilatos GD: Design and construction of an environmental SEM. 4,* Scanning *12(1):23–27, 1990; Danilatos GD, Postle R: Design and construction of an atmospheric or environmental SEM. 2,* Micron *14(1):41–52; 1983; http://www.danilatos.com/public_images.htm.*

distribution, which means that for each electron, there is the probability of no scattering event occurring or for one or more scattering events to occur. Thus, there is a fraction of electrons that are transferred to the surface of the sample without suffering any collision. The density of these electrons depends on the physical properties of the gas (gas type, pressure, temperature, and velocity), working distance, geometry of the aperture, distance from the aperture to the sample, and distribution of electrons along the beam's diameter.

Since a part of the beam hits the sample without undergoing scattering, it is with that part of the beam that the image will be formed. The electrons that have undergone some scattering hit the sample far from the point of impact of the nonscattered beam, at distances that are some orders of magnitude larger than the diameter of the nonscattered beam. In this manner, the contrast will depend more on the efficiency of the detection system than on the presence of gas, although a decrease in the beam's nonscattered intensity implies a loss of contrast. For this reason, the ESEM needs to employ the electron optic technology to take advantage of this microscopy technique. In general, a loss of contrast can in practice be compensated for by other benefits, such as contrast enhancement provided by natural exposure of the sample surface, in which case the ESEM does not require a conductive coating.

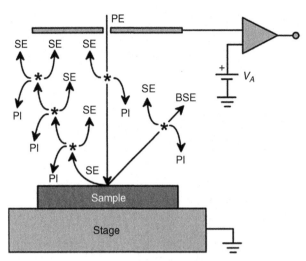

FIGURE 1.16 Operating diagram for capturing electrons in an ESEM. *ESEM,* Environmental scanning electron microscope. *Adapted from Morgan SW, Phillips MR: High bandwidth secondary electron detection in variable pressure scanning electron microscopy using a Frisch grid,* J Phys D Appl Phys *41(5):055504, 2008.*

The gas detection device (GDD) uses a positively polarized electrode (anode) (+30–600 V) located just above the sample, which collects the electrons emitted by the sample and gaseous environment between the sample and the detector (Fig. 1.16). The SEs emitted by the sample are accelerated and multiplied through the gas (usually water vapor) in the direction of the anode. As the SEs acquire sufficient kinetic energy, they ionize the gas molecules, producing positive ions. The positive ions move away from the anode toward the "sample-carrying stage" where they can neutralize the charge buildup effects of the primary electron beam. The proportion of electrons collected by the GDD coming from the sample surface or from the environment depends on the working distance, gas pressure in the chamber (P1 pressure), electron beam acceleration voltage, and collector voltage of the detector.

The use of water vapor as the ambient gas in the chamber enables the stabilization of liquid water at a pressure equal to the water vapor pressure. At a specific vapor pressure, the sample's temperature should be adjusted to maintain, evaporate, or condense the water. Wet samples should be analyzed at stable vapor pressures. If the signal is obtained at a certain pressure, the sample's temperature can be used to stabilize the water.

The possibility provided by ESEM for obtaining sample images without preparation and at high gas pressure makes it possible to obtain images of dynamic processes under ambient conditions. Crystallization in an aqueous solution can be performed using a cold stage to regulate water condensation or from a solid phase using a hot stage for melting. It is thus possible to monitor the phase changes in situ using very high temperature ranges, on the order of 1000°C (Fig. 1.17) (Zucchelli et al., 2012) through the investigation of the morphology of the plant in a humid atmosphere and under variable pressure (Fig. 1.18) (Stabentheiner et al., 2010). It is a tool with great potential for use in medical sciences (Sjong, 2009).

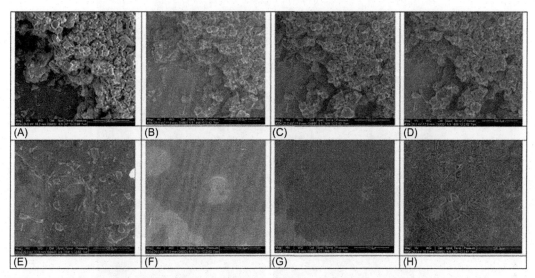

FIGURE 1.17 In situ image from an ESEM in a study on morphological changes in enamel during heating using temperatures between 57 and 900°C. (A) T = 57°C; (B) T = 499°C; (C) T = 565°C; (D) T = 608°C; (E) T = 709°C; (F) T = 791°C; (G) T = 899°C; (H) T = 900°C after 10° isotherm. *ESEM*, Environmental scanning electron microscope. *Adapted from Zucchelli A, Dignatici M, Montorsi M, Carlotti R, Siligardi C: Characterization of vitreous enamel–steel interface by using hot stage ESEM and nano-indentation techniques,* J Eur Ceram Soc *32(10):2243–2251, 2012.*

Some techniques may be expanded using ESEM, such as extension devices, fatigue control, abrasion, and other nanomechanical testing (Haque and Saif, 2002); nanogrowth and nanoclustering control in environmental conditions (Barrere et al., 2004); observation of biological nanograins (Blennow et al., 2003); studies involving liquid microinjectors for deposition or chemical reaction (Luginbuhl et al., 2000; Wei et al., 2003); and dynamic experiments under different environmental conditions (Wei et al., 2007).

7 Electron Backscatter Diffraction

Electron backscatter diffraction (EBSD) is a technique used to obtain the structural organization (crystallographic data) of the material. The EBSD's image is based on the Kikuchi diffraction patterns obtained in a transmission electron microscope (Dingley, 1984; Joy and Booker, 1971; Nishikawa and Kikuchi, 1928a, 1928b; Venables and Harland, 1973) and the introduction of the Hough transform for interpretation of the obtained images (Lassen, 1996; Lassen et al., 1992; Schmidt et al., 1991).

To perform the EBSD experiment, the sample must have a highly flat and well-polished surface, and the electron beam must hit at a grazing angle to the sample, usually 20 degrees, so the sample-carrying stage should be inclined at 70 degrees (Fig. 1.19). Using an acceleration voltage of 10–30 kV and an incident current of 1–50 nA, diffraction occurs at the point of incidence of the electron beam on the sample. With a stationary beam, the EBSD pattern emanates spherically from this point. When the primary beam interacts

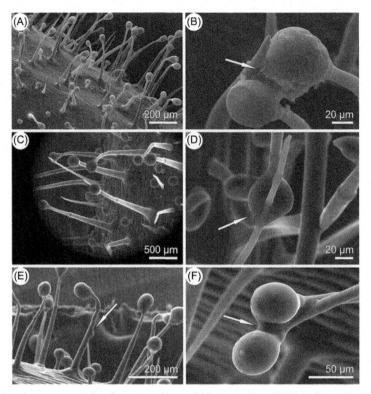

FIGURE 1.18 (A–B) Trichomes on pedicles of geranium (*P. zonale*) in conventional SEM; (C–D) wet ESEM, and (E–F) low-vacuum ESEM. The surface is characterized by nonglandular and two types of capitate trichomes (short and long; A, C, E); head cells of long capitate trichomes are shown on the right side (B, D, F). Conventional SEM: (A) overview; (B) two head cells and a nonglandular trichome are connected by residues of secretion products *(arrow)*. (C) Wet ESEM: imaging at low magnification is influenced by a limited field of view due to PLAs; (D) Recently secreted droplet *(arrow)* with a smooth appearance obscuring further surface details. (E) Low-vacuum ESEM: droplets of secretion products *(arrow)* at low magnification; (F) secretion droplets dehydrate. *Adapted from Stabentheiner E, Zankel A, Poelt P: Environmental scanning electron microscopy (ESEM)—a versatile tool in studying plants,* Protoplasma *246(1–4):89–99, 2010.*

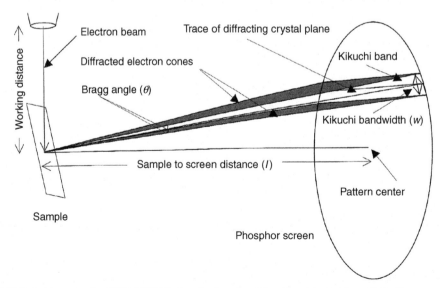

FIGURE 1.19 Detection setup for EBSD. *EBSD,* Electron backscatter diffraction. *Adapted from Oxford Instruments Analytical.* Electron backscattered diffraction explained—technical briefing.

with the crystal lattice, the electrons that have backscattered with little loss of energy are channeled and made to cross different paths that cause constructive and destructive interference. With a phosphorous screen placed at a short distance from the inclined sample, we can observe a diffraction pattern in the path of the diffracted electrons (Wells, 1999). The phosphorous screen converts the diffracted electrons into light, which is detected by a charge-coupled device (CCD)–type camera. The EBSD image detected by the CCD camera can be recorded and analyzed. It is uniquely defined by the spatial orientation of the crystal in the sample, the wavelength of the incident electron beam (which depends on the beam's acceleration), and the distance between the phosphorous screen and the sample.

This technique allows structural characteristics of the samples to be obtained. The location of the more intense Kikuchi bands (Fig. 1.20) can be clearly identified within the Hough space by the brightest peaks, while other peaks are more subtle and almost unnoticeable. It is the purpose of the peak detection algorithm to segment the most intense peaks (Fig. 1.20) and disregard the possible false peaks. To analyze the results, a line corresponding to each peak of the previous image is superimposed on the original image of the Kikuchi diffraction pattern. The lines and peaks are coded to illustrate their relationship for each point or set of points (Fig. 1.20) (Oxford Instruments Analytical).

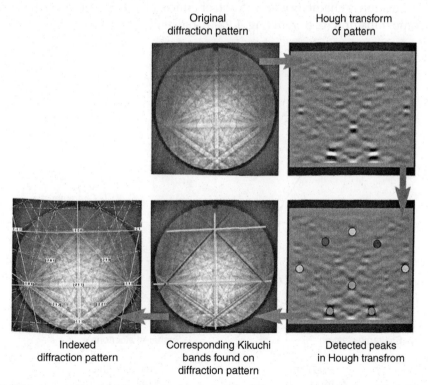

FIGURE 1.20 **EBSD sequence for obtaining structural characteristics.** This sequence includes the original diffraction pattern, Hough transformed pattern, peaks detected in the Hough transformation, corresponding Kikuchi bands found in the diffraction pattern, and indexed diffraction patterns. *EBSD*, Electron backscatter diffraction. *Adapted from Oxford Instruments Analytical.* Electron backscattered diffraction explained—technical briefing.

When we obtain the structural characteristics of a set of points within an image, it is possible to make a map of the structural characteristics of the entire image, identifying points with the same structural features.

8 Energy-Dispersive X-Ray Spectroscopy in Scanning Electron Microscopy

After hitting the sample, the electrons of the primary beam traverse a path through the sample (Fig. 1.21). Along this path, the electron can undergo interactions that generate X-rays without colliding directly with another electron. This type of interaction gives rise to what we call Bremsstrahlung X-ray radiation. In this case, the electron is scattered with little loss of energy. Otherwise, the electrons of the primary beam can collide directly with another electron and, if there is enough energy, remove the electron from its orbit. This expulsion results in the appearance of an electronic vacancy, which is filled by an electron from an outer electronic layer and of higher energy. During this transition to a more internal layer of lower energy, the electron loses energy by emitting an X-ray, which we call characteristic X-ray radiation. The energy value of the X-ray emitted during this electronic transition is characteristic for each chemical element, electronic layer (K, L, M), and electronic transition (α, β, χ, δ) (Fig. 1.22) (Goldstein et al., 2007).

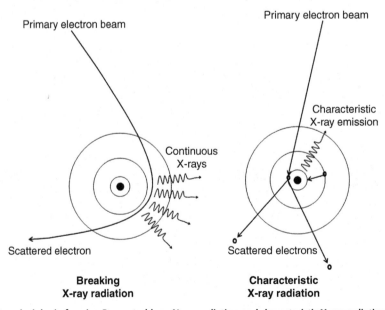

FIGURE 1.21 The principles in forming Bremsstrahlung X-ray radiation and characteristic X-ray radiation (Goldstein et al., 2007).

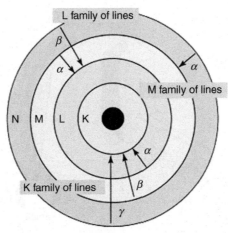

FIGURE 1.22 Families K, L, and M of characteristic X-rays. *Adapted from http://lli155-94.members.linode.com/myscope/ sem/practice/principles/gun.php; Goldstein J, Newbury DE, Joy DC, Lyman CE, Echlin P, Lifshin E, Sawyer L, Michael JR: Scanning electron microscopy and X-ray microanalysis, New York, 2007, Springer.*

The analysis of this characteristic energy enables the chemical characterization of the sample. The energy of this characteristic X-ray is captured by a dispersive X-ray energy detector.

The dispersive spectroscopy technique has been used for characterization and chemical microanalysis in conjunction with the observation of the surface morphology of samples analyzed in an SEM. This analysis technique is known as microanalysis since it performs chemical analysis on a small portion of the sample (a few micrometers) providing local chemical composition data. The lateral resolution of this analysis is directly related to the bulk of the interaction between the beam and the sample since the scattered X-rays are generated along a volume in the vicinity of the beam's impact point on the sample. The depth range of the chemical microanalysis depends on the energy of the beam that reaches the sample. Therefore, samples that are homogeneous throughout their depth can be analyzed without concern for the beam's energy. In the case of chemical microanalysis of thin films, the beam's energy greatly changes the result of the chemical microanalysis (Boyes, 2002).

Two settings are used to qualify and quantify the characteristic X-rays emitted in the sample. EDS classifies the X-rays according to their energy and simultaneously measures the entire energy spectrum, while wavelength-dispersive X-ray spectroscopy (WDS) classifies the X-rays according to their wavelength. The WDS uses a diffraction system as a means to separate the X-rays with different wavelengths. Only X-rays with a given wavelength will reach the detector for each crystal and each predetermined inclination θ (Fig. 1.23).

The main difference between the two techniques is the energy resolution. An MnKα X-ray line has a width of 135–150 eV in the EDS system, while in the WDS system, it has a width of approximately 10 eV. Fig. 1.24 shows the energy resolution of both EDS and WDS for an AuPtNb alloy.

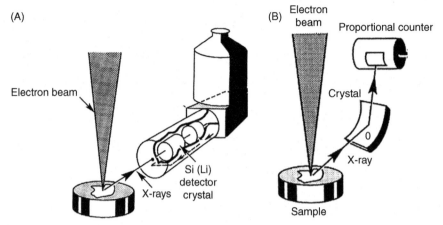

FIGURE 1.23 Detection settings for energy-dispersive X-ray spectroscopy using energy (A) and wavelength (B). *Adapted from Goldstein J, Newbury DE, Joy DC, Lyman CE, Echlin P, Lifshin E, Sawyer L, Michael JR: Scanning electron microscopy and X-ray microanalysis, New York, 2007, Springer.*

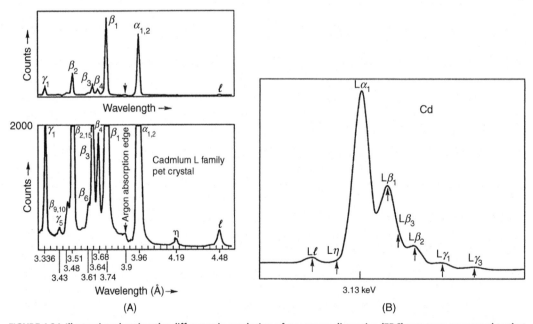

FIGURE 1.24 Illustration showing the difference in resolution of an energy-dispersive (EDS) spectrum compared to that of a wavelength-dispersive (WDS) spectrum, with both taken from an Cd L-family lines. (A) WDS X-ray spectroscopy; (B) EDS X-ray spectroscopy. *Adapted from Goldstein J, Newbury DE, Joy DC, Lyman CE, Echlin P, Lifshin E, Sawyer L, Michael JR: Scanning electron microscopy and X-ray microanalysis, New York, 2007, Springer.*

The chemical microanalysis of the light elements O, N, and C is hampered by the Be window present in most EDS-type detectors, which absorbs the low-energy X-rays emitted from these elements, preventing them from being quantified. Some types of detectors have an ultrathin window and others do not have a window, which facilitates the chemical microanalysis of these elements with low atomic numbers (Osan et al., 2000; Ro et al., 1999). EDS can identify elements present in the sample beginning at a weight percentage of approximately 0.1%, with this value being the technique's detection limit (Kuisma-Kursula, 2000).

Some new settings are proposed for analysis at the nanometer level, where the SEM has a nanomanipulator and an ion beam for surface cleaning, avoiding its contamination by hydrocarbons present on the surface of the sample (El-Gomati et al., 2011).

9 Electron Beam Lithography

Basically, EBL consists of a focused electron beam that can be deflected and interrupted through programmable commands, allowing it to be used to draw a given structure. The exposure of the resist to the electron beam changes its chemical characteristics so that, subsequently, it is possible to selectively remove the exposed material, thereby creating the desired pattern. Fig. 1.25 shows the technique of EBL. The transfer to the resist, which we call the pattern, is in direct contact with the substrate and can be used as a mask for any subsequent manufacturing process.

The resolution of EBL is related to the interaction and scattering of electrons in the resist layer and adjacent substrate. Thus, the beam's energy, beam's diameter, electron range, electron dose, type of resist, resist's thickness, and type of substrate, among other factors, must be optimized for adequate resolution in EBL. Furthermore, one should understand concepts related to exposure effects that depend on the type of pattern, known as internal and external proximity effects. Some properties of the resist, such as the sensitivity, resolution, contrast, and resistance to the bombardment, should be considered (Silva, 1996).

By optimizing these parameters, it is possible to manufacture nanoscale patterns on substrates of gallium arsenide (GaAs) (Silva et al., 1994a, 1994b) and glass (Silva et al., 1994a, 1994b), attaining resolutions on the order of 30 nm (Silva, 1996). In this manner, it is possible to draw any type of nanoscale-dimensional structure for performing various trials. In modern electron microscopes, it is possible to make a one-electron transistor in silicon (Si) (Namatsu et al., 2003).

The most used resists for EBL are polymers, particularly polymethyl methacrylate (PMMA); however, some resists can be prepared by the *sol–gel* process, including SiO_2-based resists (Silva et al., 1995) and TiO_2-based resists, showing resolutions less than 10 nm (Saifullah et al., 2003), whereas hydrogen silsesquioxane resists can reach dimensions less than 5 nm (Fig. 1.26) (Yang et al., 2009).

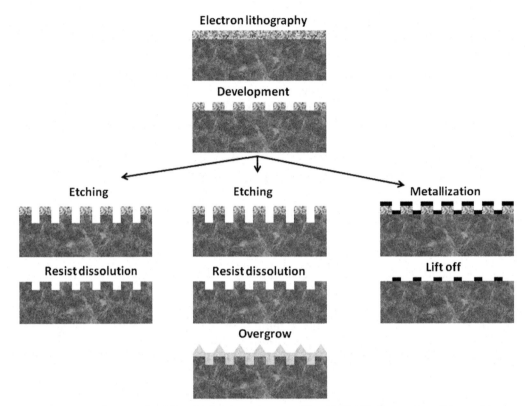

FIGURE 1.25 Process diagram using EBL and subsequent processes (Silva, 1996). *EBL,* Electron beam lithography.

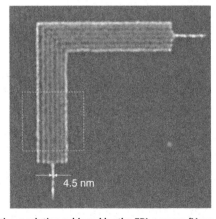

FIGURE 1.26 SEM image showing the resolution achieved by the EBL process (Yang et al., 2009). *EBL,* Electron beam lithography; *SEM,* scanning electron microscope.

10 Nanomanipulation

There is a growing and diverse range of products and devices in strategic areas, such as nanoelectronics, nanotechnology, and biotechnology, which require new tools for three-dimensional manipulation, assembly, tests, and characterization with fundamental building blocks, such as nanotubes and nanowires (Fahlbusch et al., 2005; Ru and To, 2012; Suga et al., 2009).

Scanning electron microscopy allows researchers to look into the nanometer scale, and, thus, some trials involving nanomanipulation can be performed using an SEM as an auxiliary tool to guide the researcher during the trial (Fahlbusch et al., 2005; Ru and To, 2012; Yu et al., 2002).

The main challenge in nanomanipulation is the integration between bottom-up processes, using physical and chemical techniques to obtain structures at molecular and even atomic levels and top-down processes, such as microfabrication, which aims to manufacture increasingly smaller functional structures. Fig. 1.27 shows the scale of objects for nanomanipulation (Fukuda et al., 2009).

Nanomanipulation became possible due to the evolution of nanopositioning technologies involving visual and force sensors (scanning electron microscopy and atomic force microscopy), which enabled the execution of accurate nanoscale movements (Sun et al., 2008).

The nanomanipulation technique allows the pushing, pulling, and cutting of materials using probes or microtweezers guided by piezoelectric tube actuators, thin ferroelectric

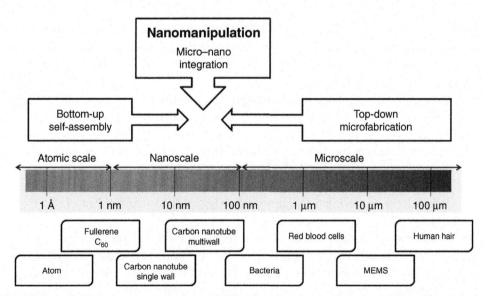

FIGURE 1.27 Scale of objects for nanomanipulation. *MEMS*, Microelectromechanical system. *Adapted from Fukuda T, Nakajima M, Liu P, Elshimy H: Nanofabrication, nanoinstrumentation and nanoassembly by nanorobotic manipulation, Int J Robotics Res 28(4):537–547, 2009.*

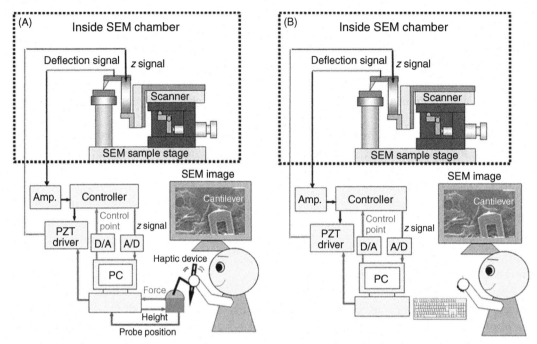

FIGURE 1.28 (A) Nanomanipulation scheme using the teleoperation mode; (B) nanomanipulation scheme using the automatic operation mode. *SEM*, Scanning electron microscope. *Adapted from Iwata F, Mizuguchi Y, Ko H, Ushiki T: Nanomanipulation of biological samples using a compact atomic force microscope under scanning electron microscope observation,* J Electron Microsc *60(6):359–366, 2011.*

films, or microelectromechanical systems (MEMS) actuated by electrical, thermal, or capacitive principles. The designs of these actuators consider their accuracy, range of motion, degree of freedom, response time, and linearity (Sitti, 2001; Sul et al., 2006; Woo et al., 2012; Zhang et al., 2013).

The nanomanipulators allow manipulation in a controlled manner, for example, by dragging along a surface or gripping and moving to a new position. The nanomanipulation control can use both the teleoperation mode and the automatic mode. In the teleoperation mode, the operator manipulates the objects directly using a man–machine interface, in which the operator participates in the feedback (visual and strength) of the experiment (Fig. 1.28A). In automatic nanomanipulation, the operator controls nanorobots sending commands for them to perform the operation (Fig. 1.28B) (Guthold et al., 2000; Iwata et al., 2011; Ru and To, 2012; Zhang et al., 2013).

Thanks to advances in nanomanipulation techniques, it will be possible to integrate various types of nanoscale materials, giving rise to new and very sensitive devices.

Fig. 1.29 shows the placement of a nanowire on a MEMS structure using two probes to drag the nanowire over the surface into the position where one wishes to fixate it. This fixation on the MEMS structure is performed through nanosoldering using a deposition process induced by an electron beam (Fukuda et al., 2009; Ye, 2012).

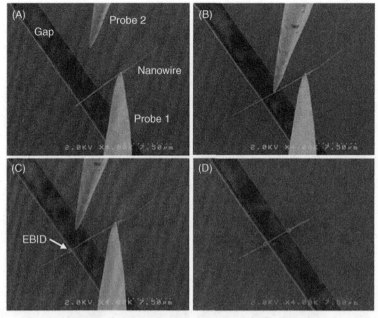

FIGURE 1.29 Nanowire placement. (A) Probe 1 positions the nanowire over the gap of the device, followed by the probe's landing on the device. (B) Probe 2 pushes the nanowire into contact with the device. (C) The nanowire is EBID-soldered to the left edge of the gap, followed by the retraction of probe 2. (D) The nanowire is EBID-soldered to the right edge of the gap, followed by the electrical breakdown of the nanowire between the second EBID point and probe 1 for the retraction of probe 1. *Adapted from Ye X:* Towards Automated Nanomanipulation Under Scanning Electron Microscopy, *Thesis submitted in conformity with the requirements for the degree of Master of Applied Science, University of Toronto, 2012, Graduate Department of Mechanical and Industrial Engineering.*

The microtweezers are thermoelectrically driven and can achieve high forces on a small scale, allowing the operator to choose the position of a nanostructure to grab and remove it from the substrate it is attached to and then position this nanostructure in a chosen location. This ability enables the nanoassembly of nanostructured devices and is a challenge for creating prototypes and sensors at the nanoscale (Chronis and Lee, 2005; Vidyaa and Arumaikkannu, 2011; Williams et al., 2002; http://www.comsol.com/paper/download/37335/Sardan.pdf). Fig. 1.30 shows a microtweezer positioning itself to touch, push, close, open, and grab a nanowire grown on a surface (Molhave et al., 2006).

One of the areas where nanomanipulation may act will be the investigation of the properties of biological materials using the conditions offered by an ESEM. The use of buckling on a nanorod as a force sensor for characterizing the stiffness of individual cells may be used in the future (Fig. 1.31). This method of detecting stiffness variations of individual cells may be applied for the diagnosis of diseases in the future based on the analysis of the stiffness of a single cell (Ahmad et al., 2011).

Fig. 1.32 shows a mechanical tensile test being carried out on nanorods with the help of an SEM for visualization (adapted from Kaplan-Ashiri and Tenne, 2007).

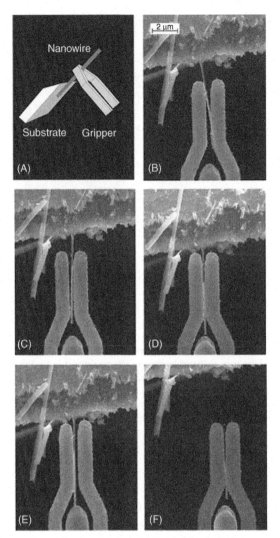

FIGURE 1.30 Positioning tweezers to nanomanipulate a nanowire. (A) Setup; (B) touch; (C) push; (D) close; (E) open; (F) grip. *Adapted from Molhave K, Wich T, Kortschack A, Boggild P: Pick-and-place nanomanipulation using microfabricated grippers,* Nanotechnology *17:2434–2441, 2006.*

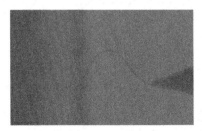

FIGURE 1.31 The large deformation of nanotubes in the postbuckling phase. *Adapted from Kaplan-Ashiri I, Tenne R: Mechanical properties of WS2 nanotubes,* J Cluster Sci *18(3):549, 2007.*

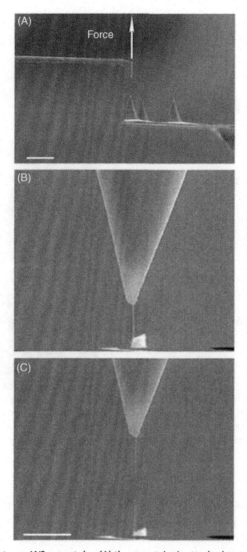

FIGURE 1.32 SEM in a tensile test on a WS₂ nanotube: (A) the nanotube is attached to two ends of a cantilever; (B) a force is applied to move one of the cantilevers away; and (C) the nanotube breaks. *SEM, Scanning electron microscope. Adapted from Kaplan-Ashiri I, Tenne R: Mechanical properties of WS2 nanotubes, J Cluster Sci 18(3):549, 2007.*

Fig. 1.33 shows an experiment where an SEM is used to guide an experiment involving the capture, transport, release, and dragging of an Fe_2O_3 nanoparticle, aided by a nanomanipulator comprising a tungsten tip on which a carbon nanotube is attached (Suga et al., 2009).

Although great progress has been made, challenges are still present. To understand the complex interactions on the nanometer scale, scientists will always be working toward improving the technologies of handling, transporting, classifying, and integrating samples in different environments (Fatikow and Eichhorn, 2008; Ru et al., 2012; Zheng, 2013).

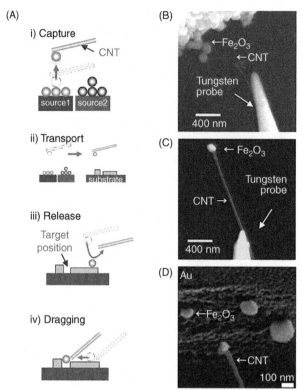

FIGURE 1.33 (A) Schematic of the nanomanipulation procedure. SEM images of (B) the CNT probe and NPs during the capture step, (C) a Fe_2O_3 NP on top of the CNT probe during the transport step, and (D) a NP pushed by the CNT probe during the dragging step. *Adapted from Suga H, Naitoh Y, Tanaka M, Horikawa M, Kobori H, Shimizu T: Nanomanipulation of single nanoparticle using a carbon nanotube probe in a scanning electron microscope,* Appl Phys Express *2(5), 2009.*

As stated, the nanomanipulation systems can be used to change, interact with, and control phenomena in a precise manner on the nanoscale. Nanomanipulation has important future applications in various areas. The use of precise and localized nanomanipulation of biological objects, such as deoxyribonucleic acid (DNA), ribosomes, proteins, and enzymes, can help develop new testing methods not yet achieved by current traditional methods (Shen et al., 2011). Computer data storage technologies may develop new types of tests, and the technologies of object assembly on the nanoscale will enable the development of complex machines composed of precisely positioned and assembled parts involving molecules, nanotubes, nanorods, and nanoparticles (Chang et al., 2011; Jeon et al., 2010). These nanomachines will need to be tested and characterized according to their properties on the nanoscale, such as friction, adhesion, and electrical and mechanical properties, which will guide their operational aspects (Acosta et al., 2011).

As it is a novel area, nanomanipulation presents some challenges to be overcome at present and certainly others will emerge along the way. Some challenges include the following: fabricating nanostructures assembled with precision and repeatability; modeling nanoforces in the air and in liquids, particularly electrostatic forces and nanotribology; and tridimensional assembly using nanotweezers and involving physical and chemical modeling of the three-dimensional nanostructure (Sitti, 2001). The preparation of nanomanipulated samples requires experience with adhesion, chemistry, nanotribology, and materials science to obtain standards for nanomanipulation procedures. Nanomanipulation systems need to be developed to obtain a reliable feedback to be able to perform accurate and stable experiments as well as provide an interpretation of the forces on the nanoscale (Sitti, 2001).

List of Symbols

δ Yield of backscattered electrons
η Yield of secondary electrons
σ Yield of emitted electrons
nie   Number of incident electrons
nse   Number of secondary electrons
nbe   Number of backscattered electrons
P0   Pressure at the gas entrance of the ESEM
P1   Pressure at the sample chamber of the ESEM
P2   Pressure at the first stage of the column of the ESEM
P3   Pressure at the second stage of the column (electron gun) of the ESEM

References

Acosta JC, Hwang G, Polesel-Maris J, Regnier S: A tuning fork based wide range mechanical characterization tool with nanorobotic manipulators inside a scanning electron microscope, *Rev Sci Instrum* 82(3):035116, 2011.

Ahmad MR, Nakajima M, Fukuda T, Kojima S, Homma M: Single cells electrical characterizations using nanoprobe via ESEM nanomanipulator system. In *9th IEEE conference on nanotechnology, IEEE-NANO*, 2009, pp 589–592.

Ahmad MR, Nakajima M, Kojima S, Homma M, Fukuda T: Buckling nanoneedle for characterizing single cells mechanics inside environmental SEM, *IEEE Trans Nanotechnol* 10(2):226–236, 2011.

Ahmad MR, Nakajima M, Kojima M, Kojima S, Homma M, Fukuda T: Instantaneous and quantitative single cells viability determination using dual nanoprobe inside ESEM, *IEEE Trans Nanotechnol* 11(2):298–306, 2012.

Asahina S, Uno S, Suga M, Stevens SM, Klingstedt M, Okano Y, Kudo M, Schueth F, Anderson MW, Adschiri T, Terasaki O: A new HRSEM approach to observe fine structures of novel nanostructured materials, *Microporous Mesoporous Mater* 146(1–3):11–17, 2011.

Barrere F, Snel MME, Van Blitterswijk CA, De Groot K, Layrolle P: Nano-scale study of the nucleation and growth of calcium phosphate coating on titanium implants, *Biomaterials* 25(14):2901–2910, 2004.

Blennow A, Hansen M, Schulz A, Jorgensen K, Donald Am, Sanderson J: The molecular deposition of transgenically modified starch in the starch granule as imaged by functional microscopy, *J Struct Biol* 143(3):229–241, 2003.

Bogner A, Jouneau P-H, Thollet G, Basset D, Gauthier C: A history of scanning electron microscopy developments: towards "wet-STEM" imaging, *Micron* 38(4):390–401, 2007.

Boyes ED: High-resolution and low-voltage SEM imaging and chemical microanalysis, *Adv Mater* 10(15):1277, 1998.

Boyes ED: Analytical potential of EDS at low voltages, *Mikrochim Acta* 138:225–234, 2002.

Cazaux J: About the role of the various types of secondary electrons (SE1; SE2; SE3) on the performance of LVSEM, *J Microsc* 214(part 3):341–347, 2004.

Chang M, Chen P-C, Lin C-P, Chang YH: Assembly of nanostructures by using a mechanical nanomanipulator. In *Advanced design and manufacture III. Key engineering materials* (vol 450), 2011, pp 263–266.

Chronis N, Lee LP: Electrothermally activated SU-8 microgripper for single cell manipulation in solution, *J Microelectromech Syst* 14(4):857–863, 2005.

Delobelle B, Courvoisier F, Delobelle P: Morphology study of femtosecond laser nano-structured borosilicate glass using atomic force microscopy and scanning electron microscopy, *Opt Lasers Eng* 48(5):616–625, 2010.

Dingley DJ: Diffraction from sub-micron areas using electron backscattering in a scanning electron microscope, *Scanning Electron Microscopy* 2:569–575, 1984.

El-Gomati MM, Walker CGH, Zha X: Towards quantitative scanning electron microscopy: applications to nano-scale analysis, *Nucl Instrum Methods Phys Res A Accelerator Spectrometers Detectors Assoc Equipment* 645(1):68–73, 2011.

Erlandsen S, Chen Y, Frethem C, Detry J, Wells C: High-resolution backscatter electron imaging of colloidal gold in LVSEM, *J Microsc* 211(3):212–218, 2003.

Fahlbusch S, Mazerolle S, Breguet JM, Steinecker A, Agnus J, Perez R, Michler J: Nanomanipulation in a scanning electron microscope, *J Mater Process Technol* 167(2–3):371–382, 2005.

Fatikow S, Eichhorn V: Nanohandling automation: trends and current developments, *Proc Inst Mech Eng [C]* 222(7):1353–1369, 2008.

Fukuda T, Nakajima M, Liu P, Elshimy H: Nanofabrication, nanoinstrumentation and nanoassembly by nanorobotic manipulation, *Int J Robotics Res* 28(4):537–547, 2009.

Goldstein J, Newbury DE, Joy DC, Lyman CE, Echlin P, Lifshin E, Sawyer L, Michael JR: *Scanning electron microscopy and X-ray microanalysis*, New York, 2007, Springer.

Griffin BJ: A comparison of conventional Everhart–Thornley style and in-lens secondary electron detectors—a further variable in scanning electron microscopy, *Scanning* 33(3):162–173, 2011.

Guthold M, Falvo MR, Matthews WG, Paulson S, Washburn S, Erie DA, Superfine R, Brooks FP, Taylor RM: Controlled manipulation of molecular samples with the nanomanipulator, *IEEE-ASME Trans Mechatronics* 5(2):189–198, 2000.

Haque MA, Saif MTA: In-situ tensile testing of nano-scale specimens in SEM and TEM, *Exp Mech* 42(1):123–128, 2002.

Iwata F, Mizuguchi Y, Ko H, Ushiki T: Nanomanipulation of biological samples using a compact atomic force microscope under scanning electron microscope observation, *J Electron Microsc* 60(6):359–366, 2011.

Jeon J, Floresca HC, Kim MJ: Fabrication of complex three-dimensional nanostructures using focused ion beam and nanomanipulation, *J Vac Sci Technol B* 28(3):549–553, 2010.

Joy DC: The theory and practice of high-resolution scanning electron microscopy, *Ultramicroscope* 37(1–4):216–233, 1991.

Joy DC, Booker GR: Simultaneous display of micrograph and selected-area channeling pattern using scanning electron microscope, *J Phys E Sci Instrum* 4(11):837, 1971.

Joy DC, Joy CS: Low voltage scanning electron microscopy, *Micron* 27(3–4):247–263, 1996.

Joy DC, Pawley JB: High-resolution scanning electron-microscopy, *Ultramicroscopy* 47(1–3):80–100, 1992.

Kaminskyj SGW, Dahms TES: High spatial resolution surface imaging and analysis of fungal cells using SEM and AFM, *Micron* 39(4):349–361, 2008.

Kaplan-Ashiri I, Tenne R: Mechanical properties of WS2 nanotubes, *J Cluster Sci* 18(3):549, 2007.

Kim H, Negishi T, Kudo M, Takei H, Yasuda K: Quantitative backscattered electron imaging of field emission scanning electron microscopy for discrimination of nano-scale elements with nm-order spatial resolution, *J Electron Microsc* 59(5):379–385, 2010.

Komoda H, Yoshida M, Yamamoto Y, Iwasaki K, Nakatani I, Watanabe H, Yasutake K: Novel charge neutralization techniques applicable to wide current range of FIB processing in FIB-SEM combined system, *Microelectron Reliability* 46(12):2085–2095, 2006.

Kuisma-Kursula P: Accuracy, precision and detection limits of SEM-WDS, SEM-EDS and PIXE in the multi-elemental analysis of medieval glass, *X-Ray Spectrom* 29(1):111–118, 2000.

Lassen NCK: Automatic localization of electron backscattering pattern bands from Hough transform, *Mater Sci Technol* 12(10):837–843, 1996.

Lassen NCK, Jensen DJ, Conradsen K: Image-processing procedures for analysis of electron back scattering patterns, *Scanning Microsc* 6(1):115–121, 1992.

Leng Y: *Materials characterization—introduction to microscopic and spectroscopic method*, Hoboken, NJ, 2008, John Wiley & Sons.

Liu JY: High-resolution and low-voltage FE-SEM imaging and microanalysis in materials characterization, *Mater Characterization* 44(4–5):353–363, 2000.

Luginbuhl P, Indermuhle PF, Gretillat MA, Willemin F, De Rooij NF, Gerber D, Gervasio G, Vuilleumier JL, Twerenbold D, Duggelin M, Mathys D, Guggenheim R: Femtoliter injector for DNA mass spectrometry, *Sens Actuators B Chem* 63(3):167–177, 2000.

Molhave K, Wich T, Kortschack A, Boggild P: Pick-and-place nanomanipulation using microfabricated grippers, *Nanotechnology* 17:2434–2441, 2006.

Muray LP: Developments in low-voltage microscopy instrumentation, *Scanning* 33(3):155–161, 2011.

Namatsu H, Watanabe Y, Yamazaki K, Yamaguchi T, Nagase M, Ono Y, Fujiwara A, Horiguchi S: Fabrication of Si single-electron transistors with precise dimensions by electron-beam nanolithography, *J Vac Sci Technol* 21(1):1–5, 2003.

Nishikawa S, Kikuchi S: Diffraction of cathode rays by calcite, *Nature* 122:726–1726, 1928a.

Nishikawa S, Kikuchi S: Diffraction of cathode rays by mica, *Nature* 121:1019–1020, 1928b.

Oono N, Kasada R, Higuchi T, Sakamoto K, Nakatsuka M, Hasegawa A, Kondo S, Matsui H, Kimura A: Irradiation hardening and microstructure evolution of ion-irradiated Zr-hydride, *J Nucl Mater* 419(1–3):366–370, 2011.

Osan J, Szaloki I, Ro CU, Van Grieken R: Light element analysis of individual micro particles using thin-window EPMA, *Mikrochim Acta* 132(2–4):349–355, 2000.

Oxford Instruments Analytical. *Electron backscattered diffraction explained—technical briefing*.

Ro C, Osan J, Van Grieken R: Determination of low-Z elements in individual environmental particles using windowless EPMA, *Anal Chem* 71(8):1521–1528, 1999.

Ru C, To S: Contact detection for nanomanipulation in a scanning electron microscope, *Ultramicroscopy* 118:61–66, 2012.

Ru CH, Zhang Y, Huang HB, Chen T: An improved visual tracking method in scanning electron microscope, *Microsc Microanal* 18(3):612–620, 2012.

Saifullah MSM, Subramanian KRV, Tapley E, Kang DJ, Welland ME, Butler M: Sub-10 nm electron beam nanolithography using spin-coatable TiO2 resists, *Nano Lett* 3(11):1587–1591, 2003.

Sawyer L, Grubb DT, Meyers GF: *Polymer microscopy*, New York, 2008, Springer.

Schatten H: Low voltage high-resolution SEM (LVHRSEM) for biological structural and molecular analysis, *Micron* 42(2):175–185, 2011.

Schmidt NH, Bildesorensen JB, Jensen DJ: Band positions used for online crystallographic orientation determination from electron back scattering patterns, *Scanning Microsc* 5(3):637–643, 1991.

Seiler H: Secondary-electron emission in the scanning electron-microscope, *J Appl Phys* 54(11):R1–R18, 1983.

Shen YJ, Ahmad MR, Nakajima M, Kojima S, Homma M, Fukuda T: Evaluation of the single yeast cell's adhesion to ITO substrates with various surface energies via ESEM nanorobotic manipulation system, *IEEE Trans Nanobiosci* 10(4):217–224, 2011.

Silva MAP: *Estudo de Litografia por Feixe de Elétrons para a Produção de Padrões Sobre Substratos de Heteroestruturas Semicondutoras,* Dissertação de Mestrado, IFSC USP, 1996.

Silva MAP, Nastaushev Y, Basmaji P, Rossi JC, Aegerter MA: Geração de máscaras em escala nano e micromérica por litografia eletrônica. In *Congresso Brasileiro de Cerâmica, CBC,* Blumenal/SC, junho 1994.

Silva MAP, Basmaji P, Aegerter MA, Nastaushev YV, Gusev GM, Rossi JC: Nanolitografia por feixe de elétrons com SEM em polimetilmetacrilado depositado sobre substrato de vidro, 11 Congresso Brasileiro de Engenharia e Ciência dos Materiais CBCIMAT, Águas de São Pedro/SP, 1994.

Silva MAP, Dahmouche K, Canto LB, Aegerter MA: PMMA–SiO2 hybrid thin film on glass substrate: an application to electron beam lithography, *Acta Microscópica* 4(Suppl B):25, 1995.

Sitti M: Survey of nanomanipulation systems. In *Proceedings of the 2001 1st IEEE conference on nanotechnology,* 2001, pp 75–80.

Sjong A: Environmental SEM for medical devices—an interview with ESEM expert, Scott Robinson, *Adv Mater Process* 167(4):52–53, 2009.

Stabentheiner E, Zankel A, Poelt P: Environmental scanning electron microscopy (ESEM)—a versatile tool in studying plants, *Protoplasma* 246(1–4):89–99, 2010.

Stevens SM, et al: An appraisal of high resolution scanning electron microscopy applied to porous materials, *Jeol News* 44(1):17, 2009.

Suga H, Naitoh Y, Tanaka M, Horikawa M, Kobori H, Shimizu T: Nanomanipulation of single nanoparticle using a carbon nanotube probe in a scanning electron microscope, *Appl Phys Express* 2(5), 2009.

Sul OJ, Falvo MR, Taylor RM II, Washburn S, Superfine R: Thermally actuated untethered impact-driven locomotive microdevices, *Appl Phys Lett* 89(20):203512, 2006.

Sun L, Wang J, Rong W, Li X, Bao H: A silicon integrated micro nano-positioning XY-stage for nano-manipulation, *J Micromech Microeng* 18(12):125004, 2008.

Venables JA, Harland CJ: Electron backscattering patterns—new technique for obtaining crystallographic information in scanning electron microscope, *Philos Mag* 27(5):1193–1200, 1973.

Vidyaa V, Arumaikkannu G: Hybrid design of a polymeric electrothermal actuator for microgripper, *Int J Mech Ind Eng* 1(2):31–35, 2011.

Wei QF, Mather RR, Fotheringham AF, Yang R, Buckman J: ESEM study of oil wetting behavior of polypropylene fibres, *Oil Gas Sci Technol* 58(5):593–597, 2003.

Wei QF, Liu Y, Wang X, Huang F: Dynamic studies of polypropylene nonwovens in environmental scanning electron microscope, *Polym Test* 26(1):2–8, 2007.

Wells OC: Comparison of different models for the generation of electron backscattering patterns in the scanning electron microscope, *Scanning* 21(6):368–371, 1999.

Williams PA, Papadakis SJ, Falvo MR, Patel AM, Sinclair M, Seeger A, Helser A, Taylor RM, Washburn S, Superfine R: Controlled placement of an individual carbon nanotube onto a microelectromechanical structure, *Appl Phys Lett* 80(14):2574–2576, 2002.

Woo P, Mekuz I, Chen B: Nanomanipulation system for scanning electron microscope scanning microscopies 2012: advanced microscopy technologies for defense, homeland security, forensic, life, environmental, and industrial sciences, *Proc SPIE* 8378(83780O), 2012.

Yang JKW, Cord B, Duan H, Berggren KK, Klingfus J, Nam SW, Kim KB, Rooks MJ: Understanding of hydrogen silsesquioxane electron resist for sub-5-nm-half-pitch lithography, *J Vac Sci Technol B* 27(6):2622–2627, 2009.

Ye X: *Towards automated nanomanipulation under scanning electron microscopy*, Thesis submitted in conformity with the requirements for the degree of Master of Applied Science, Graduate Department of Mechanical and Industrial Engineering, University of Toronto, 2012.

Yu MF, Wagner GJ, Ruoff RS, Dyer MJ: Realization of parametric resonances in a nanowire mechanical system with nanomanipulation inside a scanning electron microscope, *Phys Rev B* 66(7):073406, 2002.

Zhang YL, Zhang Y, Ru C, Chen BK, Sun Y: A load-lock-compatible nanomanipulation system for scanning electron microscope, *IEEE-ASME Trans Mechatronics* 18(1):230–237, 2013.

Zheng HM: Using molecular tweezers to move and image nanoparticles, *Nanoscale* 5(10):4070–4078, 2013.

Zonnevylle AC, Hagen CW, Kruit P, Schmidt-Ott A: Directed assembly of nano-particles with the help of charge patterns created with scanning electron microscope, *Microelectron Eng* 86(4–6):803–805, 2009.

Zucchelli A, Dignatici M, Montorsi M, Carlotti R, Siligardi C: Characterization of vitreous enamel–steel interface by using hot stage ESEM and nano-indentation techniques, *J Eur Ceram Soc* 32(10):2243–2251, 2012.

Further Reading

http://nanotechweb.org/cws/article/lab/37102.

www.ebsd-image.org/documentation/reference/ops/hough/op/houghtransform.html.

www.iei.liu.se/kmt/education/tmhl17-intranet/ebsd-materials/1.305569/TMHL17_EBSD_2011.pdf.

Eichhorn V: *Nanorobotic handling and characterization of carbon nanotubes inside the scanning electron microscope*, Dissertation zur Erlangung des Grades eines Doktors der Naturwissenschaften, Carl von Ossietzky Univertitat, Olderburg, Germany, 2011.

Kawaguchi K, Haarberg GM, Morimitsu M: Ordered nano particles in amorphous IrO2–Ta2O5 coatings detected by SEM with low accelerated incident electrons, *Electrochemistry* 77(10):879–881, 2009.

large.stanford.edu/courses/2007/ap273/bijoor1.

Nahm SH, Jang H-S, Jeon SK, Lee HJ: Mechanical properties evaluation of nano-structured materials in scanning electron microscope, *Mater Sci Forum* 654–656(1–3):2312–2315, 2010.

2

Atomic Force Microscopy: A Powerful Tool for Electrical Characterization

Ronald Tararam*, Pâmela S. Garcia**, Daiana K. Deda**, José A. Varela*, Fábio de Lima Leite**

*MULTIDISCIPLINARY CENTER FOR THE DEVELOPMENT OF CERAMIC MATERIALS, SÃO PAULO STATE UNIVERSITY, ARARAQUARA, BRAZIL; **FEDERAL UNIVERSITY OF SÃO CARLOS, SOROCABA, BRAZIL

CHAPTER OUTLINE

1 Introduction

In the early 1980s, the microstructural analysis of materials experienced a major breakthrough with the development of scanning probe microscopy (SPM) (Gomes et al., 2001). SPM encompasses a group of entirely original analytical techniques, involving multidisciplinary knowledge and technology, with broad applicability, as it allows nanoscale spatial resolution to be determined; it is also easy to operate under ambient conditions, either in liquid medium or under vacuum, generating three-dimensional images of the surface (Jalili and Laxminarayana, 2004). SPM techniques use instruments composed primarily of a sensing probe that interacts with the surface, piezoelectric ceramics for sample positioning and scanning, feedback circuits to control the vertical position of the probe, and a computer to control scanner movements, store data, and convert data into images using specific software (Stadelmann, 2016).

The essential component of SPM is the probe, which may adopt various architectures, such as a needle-shaped metallic probe for tunneling, resulting in scanning tunneling microscopy (STM) (Mironov, 2004; Müller, 2012), or a probe attached to a flexible cantilever for force detection, which characterizes atomic force microscopy (AFM) (Stadelmann, 2016; Zavala, 2008). The probe can also consist of an optical fiber with a very narrow aperture, which characterizes scanning near-field optical microscopy (SNOM) (Digital Instruments Veeco Metrology Group, 2008; Szymanski et al., 2005).

STM, the first SPM technique, was created in 1981 as a result of the work of Gerd Binnig and Heinrich Rohrer at the IBM Laboratories in Zurich (Binning et al., 1982). It was the first instrument capable of generating images of surfaces with real atomic resolution and good conductivity. AFM was developed in 1986 by Binnig, Quate, and Gerber by modifying a scanning tunneling microscope, combining the apparatus with a profilometer (which quantifies roughness on a microscopic scale), thus enabling the measurement of forces in different types of materials for the analysis of metals and semiconductors or insulating surfaces (Binning et al., 1986).

AFM unquestionably contributed to the emergence of several other techniques of the SPM family. Many other image acquisition methods originated because of some modifications of the basic principle, including lateral force microscopy (LFM) (Arias et al., 2006; Heaton et al., 2012; Jalili and Laxminarayana, 2004), phase-contrast microscopy (phase imaging) (Babcock and Prater, 2012; Chen et al., 1998; Jalili and Laxminarayana, 2004), and force modulation microscopy (FMM) (Digital Instruments Veeco Metrology Group, 2008; Jalili and Laxminarayana, 2004) among others, thereby generating a wide range of information from techniques closely related to AFM.

In reality, the rapid development of AFM-derived techniques involves several surface properties, such as chemical interactions between functionalized probes and surfaces; mechanical properties such as friction or hardness; electrical properties such as surface potential, polarization, and spatial charges; and magnetic and thermal properties (Gomes et al., 2001; Jalili and Laxminarayana, 2004; Zavala, 2008). Furthermore, properties can be analyzed in conjunction, for example, by analyzing biological material in vivo to study biochemical interactions (Steffens et al., 2012; Webb et al., 2011), electrochemical properties with in situ characterization of corrosion phenomena (Davoodi et al., 2007; Szunerits et al., 2007), electromechanical properties to study material piezoelectricity (Kalinin et al., 2006; Kholkin et al., 2007a), and photoelectrical properties with application in solar cells (Coffey et al., 2007; Pingree et al., 2009). AFM has thus undergone several developmental stages, and due to its relative simplicity, it is a fundamentally versatile technique, with it being difficult to foresee the limits of its evolution and application to different areas of knowledge.

2 Operating Principles

Understanding how an atomic force microscope operates becomes simple by drawing an analogy with an old phonograph. The sound of a vinyl record is reproduced after the phonograph's needle passes through its grooves. In AFM, a small probe (tip) acts as said needle; while passing over a sample's surface, the probe's movement is sent to a detector, which uses software to extract the information (Digital Instruments Veeco Metrology Group, 2013).

It is essential to understand how this type of microscope operates, its basic components, and the interaction forces between the probe and the surface, as the extensive variety of AFM-derived techniques depends on the type of interaction force and its measurement.

The structure of an AFM apparatus is shown in Fig. 2.1, highlighting its basic components. The AFM measures forces between the probe and the sample through a probe deflection system, with a laser beam incident on a flexible rod called a cantilever. The probe, located at the end of the cantilever, interacts with the sample surface during the scan (Butt et al., 2005; Giessibl, 2003). The method is very sensitive to surface irregularities; changes in sample topography induce movement of the cantilever, which deflects the laser beam toward the photodetector. The changes in the photodetector output are used to adjust the movement of the piezoelectric ceramic (scanner) in the z direction, whose value

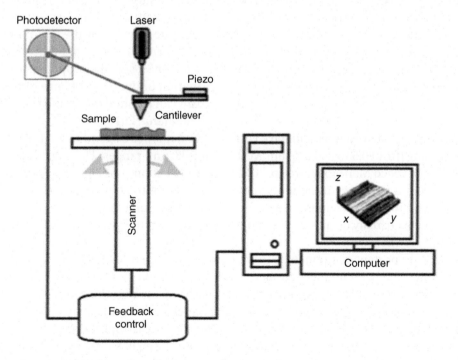

FIGURE 2.1 Basic components of the AFM system. *AFM,* Atomic force microscopy.

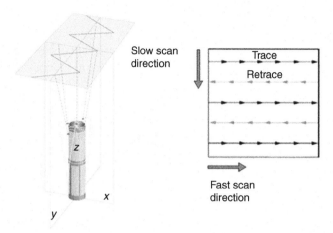

FIGURE 2.2 Raster scanning motion performed by a typical tube scanner. The voltage applied on the x- and y-axes generates the scanning pattern.

is recorded as a function of the scanning coordinates (x, y) to generate a three-dimensional image (Digital Instruments Veeco Metrology Group, 2013; Stadelmann, 2016).

The main hardware of an AFM includes the probe and the piezoelectric scanning component or scanner; their combination generates images by means of sophisticated control electronics (Cappella and Dietler, 1999). The scanning systems (scanners) enable motion control over very small distances, which is made possible by the use of piezoelectric ceramics, usually lead zirconate titanate (PZT) (Digital Instruments Veeco Metrology Group, 2013) doped with a number of substances to confer specific properties to the material. These materials experience physical deformation under the influence of an electric field. Thus, a high-precision positioning mechanism can be obtained by means of an electronic feedback loop coupled to the ceramics (Stadelmann, 2016).

Commercial microscopes can have various scanner types depending on the maximum x–y scanning (from 0.5 to 200 μm) and vertical (0.5–20 μm) ranges. The tube shape (Digital Instruments Veeco Metrology Group, 2013) is the most common (Fig. 2.2). This configuration provides a raster scanning motion, with trace and retrace pairs in the fast scan. The slow scan occurs via movement in a perpendicular direction to the fast scan. The x–y scan of a given area usually occurs at low frequency (1 Hz), resulting in a slow imaging process (Mironov, 2004).

Another important component is the conjunction of the probe and the cantilever. There is a vast diversity of commercial probes, varying in geometry and composition, and cantilevers of various lengths and spring constants. The most common probes consist of pyramidal silicon nitride (Si_3N_4) supported by triangular cantilevers and conical silicon (Si) with rectangular cantilevers (Choosing AFM Probes for Biological Applications, 2012), as shown in Fig. 2.3. Silicon probes are not as hard as those composed of silicon nitride. However, the former can be doped to become conductive, can be coated with various materials for specific applications, and are sharp to provide greater topographic detail in the obtained images.

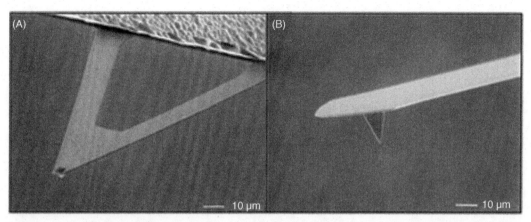

FIGURE 2.3 Scanning electron microscopy (*SEM*) images: (A) silicon nitride probe in a triangular cantilever and (B) silicon probe in a rectangular cantilever.

The characteristics of the probe are fundamental to obtaining good images, as the shape (conical or pyramidal) directly influences the contact with the sample surface (Giessibl, 2003). Said contact is related to the aspect ratio, the ratio between the height and the diameter of the probe base, and the radius of the probe tip, which is responsible for the resolution (Garcia and Perez, 2002). The larger the aspect ratio is, the finer the probe, which will allow better definition of surface irregularities.

The operation of the atomic force microscope is also related to the forces of interaction between the probe and the sample. The interaction forces between the atoms of the probe and the atoms of the sample surface are measured, and the results are computationally converted into images. These forces depend on many factors, such as the distance between the probe and the sample, the probe geometry, the presence of contaminants that will interfere with image formation, and even the compositions of the materials that constitute the probe and sample (Garcia and Perez, 2002).

On approaching the sample, the probe will be attracted by several attractive forces in the region, such as van der Waals forces (Leite et al., 2007). This attraction increases until the probe comes too close to the sample surface and atomic electron orbitals begin to repel each other, thus reducing the attractive forces with a consequent predominance of repulsive forces. The relationship of these forces with the distance between the probe and the sample surface is shown in Fig. 2.4 (Digital Instruments Veeco Metrology Group, 2008).

The following two distinct regions are highlighted: (1) contact region and (2) noncontact region. (This description is also mentioned briefly in Chapter 5 of Volume 1 of this collection, "Low-Dimensional Systems: Nanoparticles.") In the contact region, the probe is kept a few angstroms from the sample surface, where the interatomic force is repulsive. In the noncontact region, the probe is kept at a distance ranging from tens to hundreds of angstroms from the sample surface, and the force is attractive. The AFM technique explores these effects to originate different operating modes in the acquisition of topographical images (Deda et al., 2012; Hafner et al., 2001).

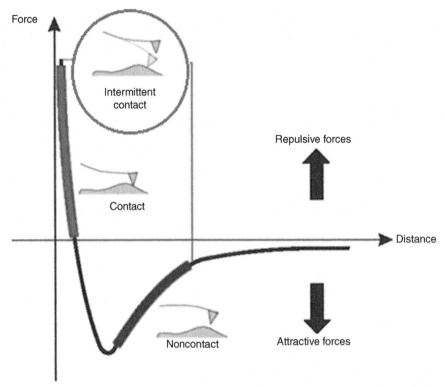

FIGURE 2.4 Schematic curve showing the dependency of the probe–sample interaction force on the interatomic distance.

One example of the application of the interaction forces between the probe and the sample is the new research area called "Nanoneurobiophysics" (Leite et al., 2015), which is presented in detail in Chapter 4 of Volume 2 of this collection. For a better understanding of the use of these techniques, it is suggested to also read "Nanosensors" (Chapter 5, Volume 2), in which the development and application of nanosensors built on AFM tips and rods are discussed (Da Silva et al., 2013, 2014; Deda et al., 2013; Garcia et al., 2015; Moraes et al., 2015; Steffens et al., 2014a, 2014b, 2014c), and "Molecular Modeling Applied to Nanobiosystems" (Chapter 7, Volume 2), which is considered an indispensable tool for understanding the in situ and in vivo application of nanosystems, including nanosensors (Amarante et al., 2014; Bueno et al., 2014; Oliveira et al., 2013).

3 Operating Modes

The different operating modes for obtaining AFM images mainly refer to the distance between the probe and the sample surface on scanning. These scan modes can be classified according to the force regimes: contact, noncontact, and intermittent contact. In the contact mode, the cantilever flexes and experiences repulsive forces, and the probe makes light "physical contact" with the sample surface. The noncontact mode operates within a very unstable region of attractive forces, and it has not found general applicability due to the adhesive forces on the surface. This limitation was circumvented with the use of the intermittent contact mode, wherein the cantilever is forced to oscillate over the sample surface in such a way that the probe periodically touches the surface; hence, the acting force in this operating mode is sometimes attractive and sometimes repulsive (Fig. 2.4) (Digital Instruments Veeco Metrology Group, 2013; Prater et al., 2012). The main operating modes, contact and intermittent contact, will be discussed in Sections 3.1 and 3.2.

3.1 Contact Mode

The contact mode is the most widely used mode of generating AFM images (Deda et al., 2012), with a higher scanning speed and good resolution (Meyer et al., 2003). In this scanning mode, the cantilever exerts a constant force on the surface (Fiorini et al., 2001). More resilient probes (silicon nitride) are frequently used, supported on a triangular cantilever (Fig. 2.3A), to prevent torsional forces, although some techniques actually make use of this information (torsional resonance mode) (Digital Instruments Veeco Metrology Group, 2008). However, depending on the sample stiffness, there is the possibility that the probe may become deformed; this can be minimized with the use of low-spring-constant cantilevers (Cappella and Dietler, 1999). Moreover, drag forces generated during scanning may damage the sample surface, which is particularly harmful to sensitive biological samples and materials weakly adhered to the substrate.

The contact between the probe and the surface occurs when the van der Waals force becomes positive (repulsive interaction force); the cantilever bends due to an elastic deformation, which can be described by Hooke's Law using the formula $F = -k \cdot x$, where k is the spring constant and x describes the vertical movement of the cantilever (Deda et al., 2012). The mechanical properties of the cantilever are determined by the spring constant, k, and the resonance frequency, ω, which will be discussed in more detail in the next section (Giessibl, 2003). Large deflections are required to achieve high sensitivity, and, thus, cantilevers with low spring constants are more appropriate.

When performed in air, the measurement may be affected by capillary forces, which occur due to the surface tension of a thin water layer present between the tip and the sample surface (Fig. 2.5) (Cappella and Dietler, 1999; Clemente and Gloystein, 2008; Giessibl, 2003; Mironov, 2004). This force is attractive and stems from the surface tension of water.

As the probe scans the surface of the sample, it passes through different points, causing the deflection of the cantilever. This deflection is measured by a laser beam reflecting off the cantilever end and reaching a photodiode. A feedback circuit maintains the cantilever deflection constant and moves the cantilever vertically for every point (Garcia and Perez, 2002).

The vertical position of the cantilever and the corresponding (x, y) position are stored in the computer, forming the topographic image of the sample. This measurement system provides high z-axis sensitivity, detecting subangstrom vertical displacements (Digital Instruments Veeco Metrology Group, 2008; Stadelmann, 2016).

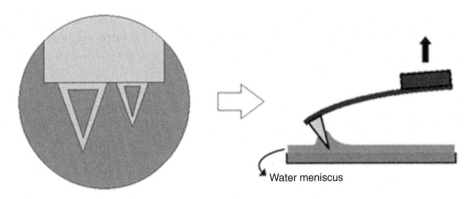

FIGURE 2.5 The contact mode and capillary adhesion force.

3.2 Intermittent Contact Mode

The intermittent contact mode was developed to overcome certain limitations found in the contact mode. The scan performed in the intermittent contact mode induces few surface alterations due to the low frictional forces applied to the sample (Schönherr and Vancso, 2010; Wawkuschewski et al., 1995). Thus, several researchers use this technique for high-resolution imaging of sensitive samples (Ares et al., 2016; Gobi et al., 2007; Gourianova et al., 2005; Kämmer et al., 2016; Smith et al., 2003) and it may be referred to in some studies as "dynamic force microscopy" or the "acoustic mode" (Lohr et al., 2007).

The control of interaction forces differs between operational regimes: by the deflection of the cantilever in the contact mode and by the oscillation of the cantilever close to the sample surface in the intermittent contact mode (Bustamante and Keller, 1995; Lee, 2002). The Z position of the scanner is adjusted to maintain a constant oscillation amplitude during scanning, which is generally performed at low speed using a feedback mechanism (Schönherr and Vancso, 2010). This low scan speed poses a disadvantage during data collection but allows for greater topographical definition and image quality.

The procedure often uses silicon probes supported by resonant rectangular cantilevers (Fig. 2.3B), whose increased stiffness allows high resonant frequencies to be achieved. Furthermore, the oscillation amplitude is set to sufficiently high values (10–100 nm) to prevent the probe from adhering to the contamination layer (capillary force) (Mironov, 2004).

The advantage of this mode is that no lateral forces are applied to the sample during scanning, justifying the preference for this technique in the study of macromolecules and system dynamics (Hafner et al., 2001). The oscillation of the cantilever occurs at (or very closely to) its resonant frequency, and its amplitude is monitored as it approaches the sample surface (Bustamante and Keller, 1995; Lohr et al., 2007).

The attractive van der Waals force becomes active when the oscillating probe is within a certain distance from the sample surface, and that force contributes to the oscillatory movement; that is, the amplitude of oscillation decreases (Fig. 2.6). This variation in amplitude during scanning depends on the sample topography.

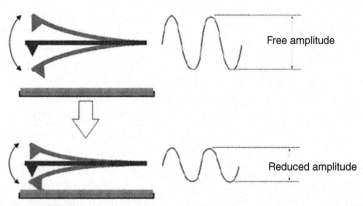

FIGURE 2.6 The intermittent contact mode and the variation in cantilever oscillation amplitude on contact with the surface.

4 Image Processing and Analysis

The AFM images are not restricted only to visual inspection but may be subjected to various processing steps, including specific algorithms and software for a more objective, quantitative, and automated characterization of the sample (Mironov, 2004).

The images captured by the system are stored as binary files containing a header with various configuration parameters and image data (Mironov, 2004; Schönherr and Vancso, 2010). The header is critical for interpreting data, providing information on the image's x, y, and z scales (Choosing AFM Probes for Biological Applications, 2012). Therefore, AFM data may initially undergo processing, followed by image analysis. Image processing involves the transformation of one image into another to reduce noise or distortions incorporated during the image capture process (Schönherr and Vancso, 2010). The software developed by the manufacturers usually contains some algorithm for calibration and correction of distortions arising from either scanner nonlinearity or interactions between the probe and topography. Additional algorithms and digital processing resources, such as filters to eliminate scratches or noise during image acquisition, can also be used (Rafael and Richard, 2002). Image analysis involves measuring the properties of elements contained in the image. Analyses involving identification, counting, and determination of the areas of objects (segmentation techniques), roughness measurements (various parameters), statistical distribution measurements (histograms), and other properties that characterize the topography of the sample are emphasized (Cremona et al., 2000; Jak et al., 2001; Meyer, 1994; Rafael and Richard, 2002).

Some commercial AFM systems usually come with software developed by the manufacturer for both data acquisition and image processing and analysis. However, the necessary resources for certain applications and measurements are not always available or have limited functionality. This leads to the development of software to perform more complex image processing and analysis functions; the majority of such software is commercially distributed, meaning that the user requires a user license. Moreover, it is necessary to check the compatibility of different image formats available on the market. As an example, one of the most complete existing commercial software packages is Scanning Probe Image Processor (SPIP), which integrates various functions and elaborate resources for the generation of three-dimensional images. The demo version is available in the reference (Image Metrology, 2012).

Concerning noncommercial software (freeware), Gwyddion (Czech Metrology Institute, 2012) and WSxM (Nanotec Electronica, 2012) are available for download. These programs possess very useful features for image processing, with several filters for image enhancement and noise reduction, Fourier transformation (Rafael and Richard, 2002), mathematical morphology (shape recognition and analysis) applied in edge identification and detection, object segmentation techniques, and analysis of surface and roughness parameters, among others. However, the three-dimensional visualization functions are limited and not very scientific.

5 Electrical Nanocharacterization

The variety of functional microdevice manufacturing methods has been growing dramatically, partly due to nanolithography techniques. However, the successful use of these devices also depends on the development of nanoscale characterization and analysis techniques, that is, nanocharacterization. Because these devices are based on electronic communication phenomena, nanoscale transport measurements are possible with SPM. In fact, after the development of AFM, which is based on short-range repulsive or attractive forces (van der Waals forces), it was possible to detect maps of different types of forces, such as long-range magnetic and electrostatic forces. Since then, the number of operating modes using the basic AFM hardware has increased rapidly, resulting in several nanocharacterization techniques that exploit electrical surface phenomena.

The enormous adaptability of commercial AFM microscopes enabled the development of accessories and software dedicated to certain electrical applications. Techniques such as electrostatic force microscopy (EFM) and scanning surface potential microscopy (SSPM) (Digital Instruments Veeco Metrology Group, 2012; Jacobs et al., 1997; Kalinin, 2002; Kalinin et al., 2005; Serry et al., 2010; Yalcin et al., 2015) are examples of modes based on the interaction of electrostatic forces available for the AFM microscope. One can also include the development of powerful tools to provide additional information on some surface electrical parameters, such as scanning spreading resistance microscopy (SSRM) (Lu et al., 2002; Suchodolskis et al., 2006; Xu et al., 2015) and conductive atomic force microscopy (c-AFM) (Kondo et al., 2015; Landau et al., 2000; Okada et al., 2002), which provide high-resolution local measurements of resistance and conductivity, aiding in the detection of defects in integrated circuits, for example (Bailon et al., 2006; Lee and Chuang, 2003). In scanning capacitance microscopy (SCM), the capacitance between the probe and the surface of the sample is measured to verify the dielectric properties of films and doping levels in semiconductors (Duhayon et al., 2002; Kopanski et al., 2000; Zavyalov et al., 1999). Piezoresponse force microscopy (PFM) (Kalinin et al., 2006) is based on the detection of the electromechanical response of the surface, enabling the observation of ferroelectric domains and measurement of local hysteresis loops (Abplanalp et al., 2001; Gruverman et al., 1997; Kalinin and Bonnell, 2002a; Kholkin et al., 2007b). Scanning impedance microscopy (SIM) is ideal for observing transport phenomena in alternating current (AC), allowing images of resistive and capacitive behavior (Kalinin and Bonnell, 2001, 2002b; Shao et al., 2003; Shin et al., 2004). The great flexibility of scanning probe microscopes, which allows significant variation in the degree of sophistication of the techniques, enables a large number of studies in the field of SPM.

Thus, it is important to classify the various AFM functional schemes for electrical characterizations to understand the imaging mechanisms. The classification of the operating regimes of the various electrical techniques is related to the interaction of the probe with the surface using different scan modes (*contact, intermittent contact,* or *noncontact*) as well as the possibility of using different forms of modulation applied to the probe or sample (*mechanical, electric,* or *magnetic modulation*). Moreover, operating regimes may be implemented in different configurations depending on the property to be measured (*one-terminal, two-terminal,* or *three-terminal*). Thus, the extensive variety of combinations enables the development of new techniques.

5.1 Classification of Operating Regimes

5.1.1 Scan Modes

During the acquisition of topographic information, the interactions between the probe and the sample surface involve van der Waals forces in attractive and repulsive regions, thus creating the three different scan modes discussed previously (contact, intermittent contact, and noncontact) (Jalili and Laxminarayana, 2004; Stadelmann, 2016; Zavala, 2008). However, some techniques explore other types of forces, such as electromagnetic interactions. For typical probe–surface distances on the order of 10–300 nm, long-range forces produced by the sample's electrostatic or magnetic fields can be detected (Stadelmann, 2016). A mechanism called LiftMode (Stadelmann, 2016) (property of the Bruker Corporation) is used for this purpose. Unlike the noncontact scan mode discussed previously, LiftMode operates in a region outside the range of attractive van der Waals forces, more than 10 nm (Fig. 2.7). The term LiftMode can also be interpreted as a noncontact scan mode in which the probe must scan the sample surface at a certain preset height to detect long-range attractive or repulsive forces (Fig. 2.8).

Topographical and local electromagnetic property images can be obtained simultaneously using LiftMode with the interleave (Digital Instruments Veeco Metrology Group, 2013) control enabled (Fig. 2.9). With the interleave control, the scanner initially performs a standard trace–retrace scan (step 1) to acquire topographic information, but it moves only half the pixel size in the direction perpendicular to the slow scan. At this stage, the scanner performs a shift in z (LiftMode) followed by a trace–retrace scan (step 2) at a constant height from the surface, thereby obtaining information on the electromagnetic forces. The first scanning step may be performed in either the contact or intermittent

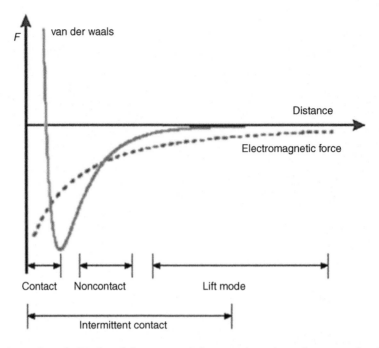

FIGURE 2.7 Dependency of van der Waals and electromagnetic forces on the probe–surface separation distance.

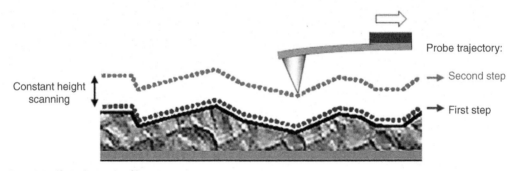

FIGURE 2.8 LiftMode scan profiles.

contact mode, whereas the second step is scanned in LiftMode, thus obtaining two images of the same region of the sample but with different information (Digital Instruments Veeco Metrology Group, 2012).

5.1.2 Modulation Schemes

The electrical interactions between the probe and the surface depend on the scanning mode employed in the analysis, which almost always involves some type of mechanical or electrical modulation applied to either the cantilever or the sample. Magnetic modulation, which will not be discussed, may be applied to the sample or to the oscillation

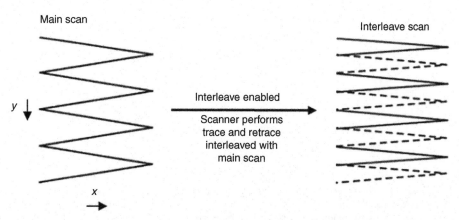

FIGURE 2.9 Description of the *x–y* scanning pattern with interleave control enabled for LiftMode operation.

of a magnetic cantilever (Florin et al., 1994; Wittborn et al., 2000). Techniques employing LiftMode use the probe as a sensor of long-range forces to measure local potentials and electrostatic fields. In this case, static measurements would be hindered by the low deflection of the cantilever, hence the need for the dynamic response of the probe by mechanical or electrical oscillation of the cantilever.

Considering mechanical modulation, an actuator (piezo; Fig. 2.1) causes the cantilever to vibrate at its resonance frequency (ω), in which the maximum oscillation amplitude is kept constant (Fig. 2.10). Information on both the cantilever oscillation amplitude and the oscillation phase may be obtained as a function of the frequency using lock-in amplifiers

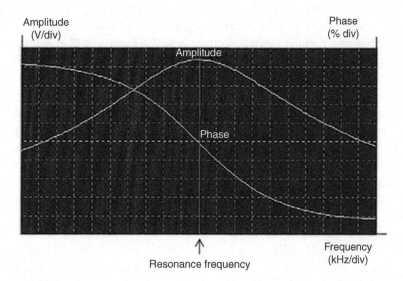

FIGURE 2.10 Typical curve exhibited by the microscope to determine the resonance frequency of the cantilever.

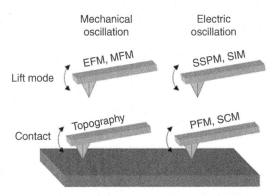

FIGURE 2.11 Operating regimes involving some electrical techniques. *EFM*, Electrostatic force microscopy; *MFM*, magnetic force microscopy; *PFM*, piezoresponse force microscopy; *SCM*, scanning capacitance microscopy; *SIM*, scanning impedance microscopy; *SSPM*, scanning surface potential microscopy.

(Serry et al., 2010). The mechanical oscillation of the cantilever close to the surface results in the intermittent contact mode, but in LiftMode, the presence of a gradient of electrostatic forces near the surface will alter the cantilever resonance frequency. Signal sensitivity is improved by detecting these variations in resonance frequency ($\Delta\omega$) during scanning. The microscope itself possesses a system to detect the resonance frequency of different cantilevers, which should be determined before starting the scan.

The number of techniques based on mechanical phenomena is not limited to cantilever oscillations; oscillations can also be applied to the sample itself, as in atomic force acoustic microscopy (AFAM) (Kalinin et al., 2005), which uses an ultrasonic transducer over the sample and can quantify surface elastic properties (Kalyan Phani et al., 2016).

An alternative approach to mechanical oscillation of the cantilever involves applying electrical modulation to the probe and cantilever, which must be conductive. In this mechanism, the actuator (piezo) is deactivated, and the probe is subjected to an AC. In LiftMode scanning, the surface electrical charges of the sample will oscillate in conjunction with the cantilever due to the alternating electrical field. SSPM uses this principle to detect the local surface potential. Other techniques, such as PFM (Kalinin et al., 2006), employ electrical modulation of the probe and usually operate in the contact scan mode to explore, for example, the electromechanical effects of a piezoelectric sample.

Electrical oscillation may also be induced in the sample to study lateral transport phenomena in AC, such as in SIM (Kalinin et al., 2005), which explores the impedance properties of materials. New techniques can be developed, but it is always convenient to classify them according to the probe–sample separation, source of oscillation, and detection (Kalinin et al., 2005), as shown in Fig. 2.11.

5.1.3 Types of Configurations

Many of the AFM electrical techniques are based on the detection of charges via surface electrostatic forces, wherein the probe acts mostly as a sensor (Kalinin et al., 2005). However, some techniques involve charge transfer between the probe and the sample, for

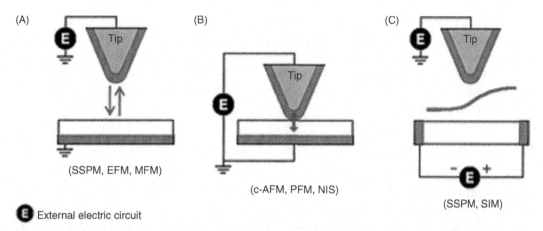

FIGURE 2.12 AFM configurations for electrical measurements: (A) *one-terminal*, (B) *two-terminal*, and (C) *three-terminal*. *AFM*, Atomic force microscopy; *c-AFM*, conductive atomic force microscopy; *EFM*, electrostatic force microscopy; *NIS*, nano impedance spectroscopy; *PFM*, piezoresponse force microscopy; *SIM*, scanning impedance microscopy; *SSPM*, scanning surface potential microscopy.

example, in c-AFM, which measures electrical current through the sample (Kalinin, 2002). Therefore, it is interesting to classify the functionality of the techniques considering the type of configuration used in the measurements, which can be *one-terminal, two-terminal,* or *three-terminal* (Kalinin, 2002; Kalinin et al., 2005).

In the *one-terminal* configuration, the probe operates solely as a sensor, detecting short-range van der Waals–type forces or long-range electromagnetic forces. The sample can be either grounded or under the influence of an external electric field. Examples of techniques using this configuration are shown in Fig. 2.12A.

In measurements employing the *two-terminal* configuration (Fig. 2.12B), the probe acts as an electrode; that is, electrical charge transfer interactions occur with a second electrode placed in the sample stage. In this case, the main scan occurs in the direct contact mode with the sample surface, exploiting either direct current properties, as in c-AFM, or AC properties, as in nanoimpedance microscopy (NIM) (Shao et al., 2003); electromechanical displacement may also be detected (PFM).

In the *three-terminal* configuration (Fig. 2.12C), two macroscopic electrodes are arranged laterally on the sample, through which voltage and current can be applied. In this case, the third terminal is the probe, which may act as a force sensor (in LiftMode) or as an electrode (in contact with the sample surface). This situation is convenient for observing and measuring lateral current transport in the sample in both AC and direct current regimes. Examples include the SSPM technique operating in LiftMode and scanning potentiometry (Trenkler et al., 1998) in the contact mode. These techniques are interesting for studying nanoscale transport phenomena in situ.

Thus, there is a vast variety of techniques for electrical measurements using the atomic force microscope, with differing applicabilities, sensitivities, and spatial resolutions. However, the main limitations of these techniques arise from the complicated geometry of the probe–surface system. In noncontact or LiftMode techniques, the capacitive force acting on the cantilever varies with the topography, thus limiting spatial resolution. The resolution in contact modes is limited by the area of effective interaction between the probe tip and the surface. The various possibilities of using AFM for the electrical nanocharacterization of surfaces result in benefits not offered by other conventional (electronic and optical) microscopy techniques.

5.2 Examples of Use

5.2.1 Electrostatic Force Microscopy

EFM (Kalinin, 2002; Kalinin et al., 2005; Serry et al., 2010; Stadelmann, 2016; Yalcin et al., 2015) is the technique most widely used to detect an electric field gradient between the probe and the sample and is therefore available in most commercial instruments. Its principle is simple; a conductive probe interacts with the sample through long-range electrostatic forces. Static charges on the surface frequently exhibit small electric fields, making it convenient to apply voltage between the probe and the sample to induce an electric field. This allows the EFM image contrast to be improved, enabling the observation of the charge distribution in addition to the estimated surface conductivity. This technique can distinguish between charges accumulated in conducting regions and charges contained in insulating regions.

EFM uses the interleave resource in LiftMode scanning (Figs. 2.8 and 2.9); that is, readings are taken in two steps for the same scan line. The first step is performed in the intermittent contact scanning mode to obtain topographic information. In the second step, the probe is raised to a certain height above the surface, and the scan is performed while maintaining the cantilever at a fixed distance from the surface. The microscopy system uses the profile of the previous scan to keep the probe–surface separation constant (Digital Instruments Veeco Metrology Group, 2012). Fig. 2.13 depicts the process in which the probe initially measures the topography (1), moves to a preset height (2), and then, in the second scan, (3) collects electric field data on the sample surface.

Fig. 2.14 shows that the principle for the detection of electrostatic forces in EFM is based on the mechanical modulation of the cantilever during the second scanning step (Kalinin et al., 2005). The presence of an electrostatic force gradient near the surface disturbs the cantilever oscillation, described by driven damped harmonic oscillator physics (Mironov, 2004). This can be understood by the modification of the effective spring constant (k_{ef}) of the oscillator, which depends on the force regime. The spring constant is smaller than the original for attractive forces ($k_{ef} < k$), as if the cantilever became "softer," decreasing its resonance frequency (Fig. 2.15A) (Digital Instruments Veeco Metrology Group, 2013). The opposite is true for repulsive forces ($k_{ef} > k$); that is, the cantilever becomes "harder," which increases the resonance frequency (Fig. 2.15B). Thus, the probe oscillates near the surface, but without touching it, and detects changes in the cantilever's resonance frequency ($\Delta\omega$). This signal is recorded with the scanning plane coordinates, yielding a 3D EFM image.

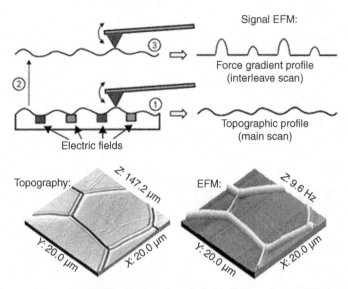

FIGURE 2.13 LiftMode scanning used in EFM for the detection of electric field gradients. *EFM*, Electrostatic force microscopy.

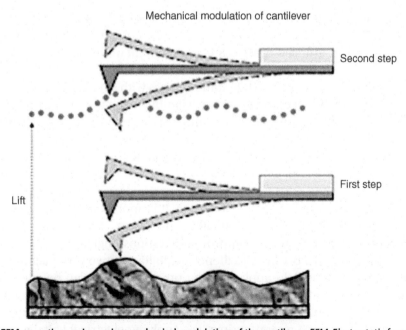

FIGURE 2.14 EFM operating regime using mechanical modulation of the cantilever. *EFM*, Electrostatic force microscopy.

(A)

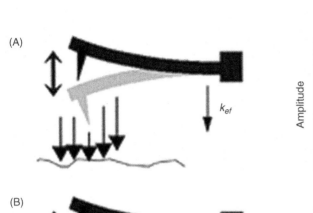
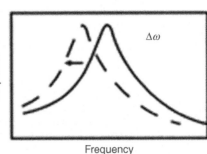

(B)

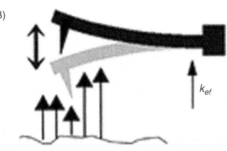
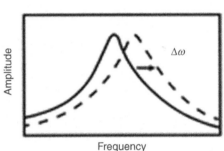

FIGURE 2.15 Comparison between (A) attractive force and (B) repulsive force effects on the cantilever spring constant (k_{ef}) and frequency (ω).

The effects shown in Fig. 2.15 generate bright and dark contrast in the EFM image related to the shift in the cantilever resonance frequency. Because of the electric field frequently applied on the probe, induced charges are predominant on the surface, creating a field of attractive forces below the baseline (dark areas in the image), regardless of the sign of the potential (positive or negative), while static charges with smaller fields contribute only weakly. This makes the EFM technique very susceptible to the topography of the sample.

Moreover, it is possible to estimate the conductivity of a sample by detecting regions of induced charge in conducting materials and confined charge in insulating materials. The physical characteristics of the cantilever will influence the intensity of the detected signal, and resonance frequencies of approximately 75 kHz are generally more suitable. The sensitivity will also depend on the coating of the conducting probe, usually Pt/Ir or Co; however, highly doped Si probes also operate well. It is appropriate that the upper region of the cantilever, where the laser beam is incident, possesses a metallic Au or Al coating to increase reflection to the detector and improve the response signal.

Because EFM detects electric field gradients, the following two variables are important: the intensity of the electric field applied to the probe and the scan height in LiftMode. Fig. 2.16A illustrates the influence of these variables on the image acquisition of a thin conductive gold layer deposited on an insulating SiO_2 substrate. The configuration type (*one-terminal*) used for EFM operation involves applying electrical voltage between the

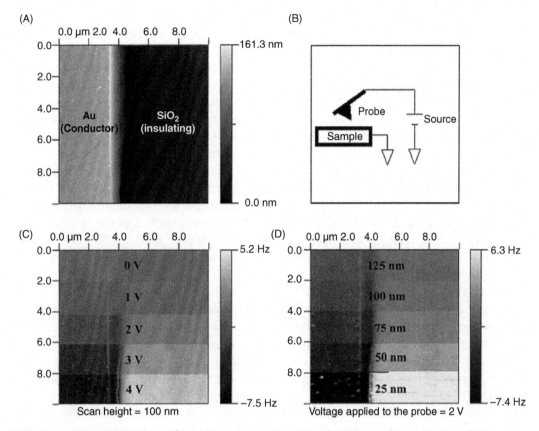

FIGURE 2.16 (A) Topographic image of a gold layer deposited on SiO₂ and (B) scheme of the application of voltage to the probe. EFM images showing the influence of the (C) voltage applied to the probe and (D) scan height. *EFM,* Electrostatic force microscopy.

probe and the sample (Fig. 2.16B), in which the probe will act solely as a force sensor. The EFM image shown in Fig. 2.16C was collected by varying the voltage applied to the probe during LiftMode scanning at a preset height of 100 nm. The image contrast becomes greater between conductive and insulating regions with increases in the voltage applied to the probe. In the conductive region (dark area), the induced electrical charges contribute with attractive forces on the oscillating cantilever, reducing its resonance frequency. A similar influence occurs with the variation in the scan height when a preset voltage is applied to the probe. In Fig. 2.16D, the image contrast increases when the probe oscillates closer to the sample surface, where electrostatic forces are more intense. It is noted that the electric field gradient over the sample surface follows a probe voltage–dependent and scan height–dependent exponential relationship, which renders EFM laborious for quantitative studies (Kalinin, 2002).

EFM was the first technique developed to investigate the electrical properties of the surfaces of materials, while also being one of the simplest techniques. Charge detection experiments began with Martin et al. in late 1987, followed by Stern et al. in 1988, and Terris et al. in 1989. Since then, EFM has become a powerful research tool in materials science.

EFM can be applied, for example, in triboelectrification, a phenomenon associated with charge transfer on contact between different materials. The only well-known contact occurs between metals from the equilibrium of the Fermi levels. However, the metal–insulator or insulator–insulator contacts are not well known, with many diverging reports in the literature (Stadelmann, 2016). Schonenberger and Alvarado (1990) deposited charges over insulating Si_3N_4 films to observe the decay of the electrostatic force signal versus time. Charge decay in this nanoscopic study occurs much faster than in macroscopic experiments on the same materials. This is further evidence that, in principle, mechanisms at the nanoscopic level are very different from those at the macroscopic level (Stadelmann, 2016).

Another important EFM field of application is the investigation of the spatial distribution of polarized charges and electric fields in ferroelectric surfaces. Saurenbach and Terris (1990) observed changes in the distribution of electrical charges at the interface of $Gd_2(MoO_4)_3$ ferroelectric domains.

Electron transport properties can also be assessed. Ehrburgerdolle and Tence (1990) observed the presence of conductive carbon islands (dark regions) dispersed in an insulating high-density polyethylene matrix (light region). Some islands were isolated, but many were connected, forming a network through which current could flow.

Recently, EFM has been used in the characterization and visualization of voltage barriers in nonlinear polycrystalline systems. Marques et al. (2007) demonstrated the possibility of using EFM to detect active potential barriers in SnO_2-based varistor ceramic samples to verify the influence of heat treatment under oxygen-rich or oxygen-poor atmospheres. The influence of dopants, which alter potential barriers in varistors, can also be evaluated using EFM (Bueno et al., 2008). Vasconcelos et al. (2006) employed the technique to identify and estimate the amount of active barriers in an SnO_2-based varistor compared with that in a commercial ZnO varistor. Gheno et al. (2007) characterized yttrium-doped barium titanate, assigning the positive temperature coefficient (PTC) effect to the formation of potential barriers at the grain boundaries, which were directly observed with EFM.

5.2.2 Scanning Surface Potential Microscopy

The first surface potential measurements using an atomic force microscope were performed by Nonnenmacher et al. (1991, 1992) who developed Kelvin microscopy [Kelvin probe force microscopy (KPFM)] for measuring the surface potentials of different metals. The basic method has undergone several improvements since then, with SSPM (Digital Instruments Veeco Metrology Group, 2012, 2013; Jacobs et al., 1997; Kalinin, 2002; Kalinin et al., 2005; Serry et al., 2010) being available in the majority of commercial microscopes. The drawbacks of EFM force gradient detection with frequency response led to the development of SSPM, which allows scanning at a preset height from the surface with high potential resolution (on the scale of millivolts) (Kalinin, 2002). Thus, the understanding of local electrical phenomena using SSPM may enhance the description of macroscopic properties.

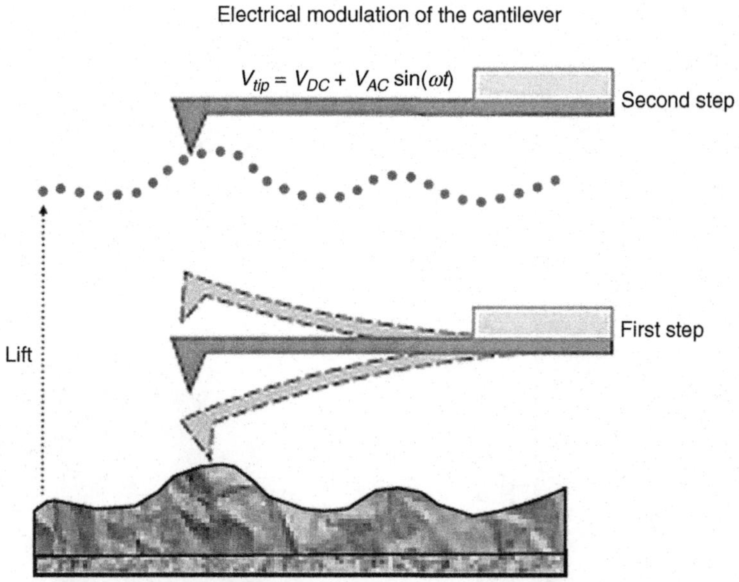

Electrical modulation of the cantilever

$$V_{tip} = V_{DC} + V_{AC} \sin(\omega t)$$

Second step

First step

Lift

FIGURE 2.17 **Electrical modulation scheme in LiftMode SSPM.** *SSPM,* Scanning surface potential microscopy.

SSPM is similar to EFM in the LiftMode scanning procedure, but differs in the excitation of the cantilever. In SSPM, an AC voltage is applied to the probe (tip), with electric modulation of the cantilever to detect electrostatic forces (Kalinin, 2002).

The mechanism for obtaining SSPM images may be described using the diagram in Fig. 2.17. The process consists of the following operations: the first trace of the principal scan is performed in the intermittent contact mode, with mechanical vibration of the cantilever to collect topographical data, and the second scan is performed in LiftMode, with cantilever vibration disabled and the application of alternating voltage (V_{AC}) directly to the cantilever and conducting probes. If there is an electric potential difference (ΔV) between the probe and the sample surface, there will be an electric driving force on the cantilever, causing it to vibrate. This allows detection of the oscillation frequency (ω) and amplitude of the cantilever. The higher the potential difference between the probe and surface is, the greater the force on the cantilever, increasing the oscillation amplitude.

If the probe and the sample have the same potential ($\Delta V = 0$), there will be no electrical force acting on the cantilever, and its oscillation amplitude will be zero. Therefore, the local surface potential is determined by adjusting the DC voltage to the probe (V_{DC}) to match the surface's electrical potential (V_{surf}), resulting in the amplitude of oscillation of the cantilever becoming zero. The DC voltage applied to the probe is recorded by the system to build a surface potential map, wherein the image contrast between light and dark is associated with positive and negative potentials, respectively.

Therefore, during an SSPM scan, two electrical components act on the probe and the cantilever, one DC and one AC. The DC component involves the adjustable voltage

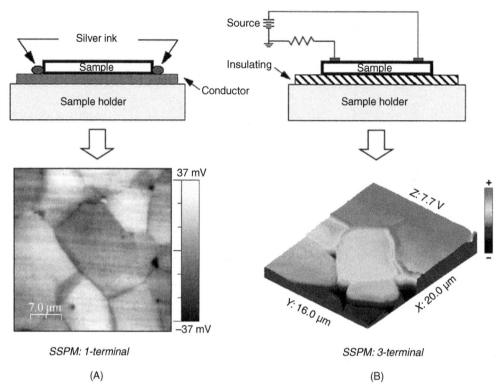

FIGURE 2.18 SSPM configuration with (A) a grounded sample or (B) lateral voltage application. *SSPM*, Scanning surface potential microscopy.

determined by the feedback system according to the surface potential. The AC component is responsible for the cantilever's electrical oscillation according to the following relationship: $V_{tip} = V_{DC} + V_{AC} \sin(\omega t)$ (Digital Instruments Veeco Metrology Group, 2012; Kalinin et al., 2005; Serry et al., 2010).

The possible SSPM operational settings may involve either the grounding of or the application of a lateral voltage to the sample (Digital Instruments Veeco Metrology Group, 2013), as shown in Fig. 2.18. In the first case, the probe operates solely as a surface potential sensor in a *one-terminal* configuration (Fig. 2.12A). The sample must be electrically connected to the metallic support of the sample stage using either conductive epoxy or silver ink. This configuration is most often used in EFM. In the second case, more suitable for SSPM, test probes from an external power source are attached to the sample with appropriate connectors so that a DC voltage applied to the electrodes can modify the surface potential. For normal operation, one must carefully ensure that there is no electrical connection between the sample and the sample stage. Thus, in situ measurements can be performed by varying the applied external lateral voltage during scanning in a typical *three-terminal* configuration (Fig. 2.12C).

Considering the main SSPM applications in nanocharacterization, we can highlight, for example, the ability to detect and quantify the contact potential differences between different materials, which is related to their work function. O'boyle et al. (1999) showed that this technique is able to distinguish components of a metallic alloy from the differences in the work function, verified by the strong contrast ascribed to intermetallic Al/Cu. One other vast application field is the characterization of the surface potential distribution in functional integrated circuits. It is possible to observe the structure of semiconductor cross-sections with *p–n* rectifying junctions and Schottky barriers to identify the depletion layer as well as variations in doped areas (Buh et al., 2000).

The detection of contaminants or defects in substrate manufacturing can be performed using SSPM. Jiang et al. (2006) reported the direct measurement of trapped charges in Si (1 1 1) crystalline defects using STM and SSPM. Electron entrapment in atomic steps and disordered domains contributed to the increase of the local work function.

Grain-boundary potential barriers in polycrystalline materials are important for the electronic properties of devices such as solar cells, gas sensors, and varistors. The spatial variation of potential barriers can be quantified using SSPM, as described by Huey and Bonnell (2000). The authors studied the in situ resistance of electrically active ZnO grain boundaries, simulating the varistor in operation. Hirose et al. (2006) studied the influence of the distribution of additives in ZnO-based varistor chips, subjecting the samples to lateral DC voltage to observe variations in the height of the potential barrier along the grains due to the current–voltage nonlinearity at the grain boundary.

List of Symbols

ω	Resonance frequency
$\Delta\omega$	Variations in the resonance frequency
F	Force
k	Spring constant
k_{ef}	Effective spring constant
ΔV	Electrical potential difference
V_{AC}	Alternating current voltage
V_{DC}	Direct current voltage
V_{sup}	Surface electrical potential
x	Vertical displacement of the cantilever

References

Abplanalp M, Barosova D, Bridenbaugh P, Erhart J, Fousek J, Gunter P, Nosek J, Sulc M: Ferroelectric domain structures in PZN–8% PT single crystals studied by scanning force microscopy, *Solid State Commun* 119(1):7–12, 2001.

Amarante AM, Oliveira GS, Bueno CC, Cunha RA, Ierich JCM, Freitas LCG, Franca EF, Oliveira ON, Leite FL: Modeling the coverage of an AFM tip by enzymes and its application in nanobiosensors, *J Mol Graph Model* 53:100–104, 2014.

Ares P, Fuentes-Perez ME, Herrero-Galan E, Valpuesta JM, Gil A, Gomez-Herrero J, Moreno-Herrero F: High resolution atomic force microscopy of double-stranded RNA, *Nanoscale* 8:11818–11826, 2016.

Arias DF, Marulanda DM, Baena AM, Devia A: Determination of friction coefficient on ZrN and TiN using lateral force microscopy (LFM), *Wear* 261(11–12):1232–1236, 2006.

Babcock KL, Prater CB: *Phase imaging: beyond topography*, disponível em: <http://www.bruker.jp/axs/nano/imgs/pdf/AN011.pdf>.

Bailon MF, Salinas PFF, Arboleda JPS: Application of conductive AFM on the electrical characterization of single-bit marginal failure, *IEEE Trans Device Mater Reliability* 6(2):186–189, 2006.

Binning G, Rohrer H, Gerber C, Weibel E: Surface studies by scanning tunneling microscopy, *Phys Rev Lett* 49(1):57–61, 1982.

Binning G, Quate CF, Gerber C: Atomic force microscope, *Phys Rev Lett* 56(9):930–933, 1986.

Bueno PR, Santos MA, Ramirez MA, Tararam R, Longo E, Varela JA: Relationship between grain-boundary capacitance and bulk shallow donors in SnO_2 polycrystalline semiconductor, *Phys Status Solidi A* 205(7):1694–1698, 2008.

Bueno CC, Amarante AM, Oliveira GS, Deda DK, Teschke O, Franca ED, Leite FL: Nanobiosensor for diclofop detection based on chemically modified AFM probes, *IEEE Sens J* 14(5):1467–1475, 2014.

Buh GH, Chung HJ, Kim CK, Yi JH, Yoon IT, Kuk Y: Imaging of a silicon pn junction under applied bias with scanning capacitance microscopy and Kelvin probe force microscopy, *Appl Phys Lett* 77(1):106–108, 2000.

Bustamante C, Keller D: Scanning force microscopy in biology, *Phys Today* 48(12):32–38, 1995.

Butt H-J, Cappella B, Kappl M: Force measurements with the atomic force microscope: technique, interpretation and applications, *Surf Sci Rep* 59(1–6):1–152, 2005.

Cappella B, Dietler G: Force–distance curves by atomic force microscopy, *Surf Sci Rep* 34(1–3):1–104, 1999.

Chen X, Davies MC, Roberts CJ, Tendler SJB, Williams PM, Davies J, Dawkes AC, Edwards JC: Interpretation of tapping mode atomic force microscopy data using amplitude–phase–distance measurements, *Ultramicroscopy* 75(3):171–181, 1998.

Choosing AFM Probes for Biological Applications—Veeco: <http://ipc.iisc.ernet.in/~ipcafm/pdf/AN44-BioTips%20Rev%20A2f.pdf>.

Clemente AV, Gloystein K: *Principles of atomic force microscopy (AFM)*, Physics of Advanced Materials Winter School, 2008.

Coffey DC, Reid OG, Rodovsky DB, Bartholomew GP, Ginger DS: Mapping local photocurrents in polymer/fullerene solar cells with photoconductive atomic force microscopy, *Nano Lett* 7(3):738–744, 2007.

Cremona M, Mauricio MHP, Carmo LCS, Prioli R, Nunes VB, Zanette SI, Caride AO, Albuquerque MP: Grain size distribution analysis in polycrystalline LiF thin films by mathematical morphology techniques on AFM images and x-ray diffraction data, *J Microsc* 197:260–267, 2000.

Czech Metrology Institute; *Gwyddion*, disponível em: <http://sourceforge.net/projects/gwyddion/files/gwyddion>.

Da Silva ACN, Deda DK, Da Roz AL, Prado RA, Carvalho CC, Viviani V, Leite FL: Nanobiosensors based on chemically modified AFM probes: a useful tool for metsulfuron-methyl detection, *Sensors* 13(2):1477–1489, 2013.

Da Silva ACN, Deda DK, Bueno CC, Moraes AS, Da Roz AL, Yamaji FM, Prado RA, Viviani V, Oliveira ON, Leite FL: Nanobiosensors exploiting specific interactions between an enzyme and herbicides in atomic force spectroscopy, *J Nanosci Nanotechnol* 14(9):6678–6684, 2014.

Davoodi A, Pan J, Leygraf C, Norgren S: Integrated AFM and SECM for in situ studies of localized corrosion of Al alloys, *Electrochim Acta* 52(27):7697–7705, 2007.

Deda DK, Bueno CC, Ribeiro GA, Moraes AS, Garcia PS, Brito B, Leite FL: Atomic force microscopy-based molecular recognition: a promising alternative to environmental contaminants detection, Méndez-Vilas A, editor: *Current microscopy contributions to advances in science and technology*, 5, Badajoz, 2012, Formatex Research Center, pp 1337–1348.

Deda DK, Pereira BBS, Bueno CC, Da Silva AN, Ribeiro GA, Amarante AM, Franca EF, Leite FL: The use of functionalized AFM tips as molecular sensors in the detection of pesticides, *Mater Res* 16(3):683–687, 2013.

Digital Instruments Veeco Metrology Group: *A practical guide to scanning probe microscopy*, 2008, disponível em: <http://www.veeco.com/pdfs/library/SPM_Guide_0829_05_166.pdf>.

Digital Instruments Veeco Metrology Group: *NanoScope IVa controller manual. Version 6.13*, disponível em: <https://depts.washington.edu/ntuf/facility/docs/NanoScope_Controller_Manual.pdf>.

Digital Instruments Veeco Metrology Group: *MultiMode SPM instruction manual*, disponível em: <http://www.cigs.unimo.it/CigsDownloads/labs/AFM2/manuali_letture/MultiMode_Manual_RevB.pdf>.

Duhayon N, Clarysse T, Eyben P, Vandervorst W, Hellemans L: Detailed study of scanning capacitance microscopy on cross-sectional and beveled junctions, *J Vac Sci Technol B* 20(2):741–746, 2002.

Ehrburgerdolle F, Tence M: Determination of the fractal dimension of carbon-black aggregates, *Carbon* 28(2–3):448–452, 1990.

Fiorini M, Mckendry R, Cooper MA, Rayment T, Abell C: Chemical force microscopy with active enzymes, *Biophys J* 80(5):2471–2476, 2001.

Florin EL, Radmacher M, Fleck B, Gaub HE: Atomic force microscope with magnetic force modulation, *Rev Sci Instrum* 65(3):639–643, 1994.

Garcia R, Perez R: Dynamic atomic force microscopy methods, *Surf Sci Rep* 47(6–8):197–301, 2002.

Garcia PS, Moreau ALD, Ierich JCM, Vig ACA, Higa AM, Oliveira GS, Abdalla FC, Hausen M, Leite FL: A nanobiosensor based on 4-hydroxyphenylpyruvate dioxygenase enzyme for mesotrione detection, *IEEE Sens J* 15(4):2106–2113, 2015.

Gheno SM, Hasegawa HL, Paulin PI: Direct observation of potential barrier behavior in yttrium–barium titanate observed by electrostatic force microscopy, *Scripta Mater* 56(6):545–548, 2007.

Giessibl FJ: Advances in atomic force microscopy, *Rev Mod Phys* 75(3):949–983, 2003.

Gobi KV, Kim SJ, Tanaka H, Shoyama Y, Miura N: Novel surface plasmon resonance (SPR) immunosensor based on monomolecular layer of physically-adsorbed ovalbumin conjugate for detection of 2,4-dichlorophenoxyacetic acid and atomic force microscopy study, *Sens Actuators B Chem* 123(1):583–593, 2007.

Gomes S, Trannoy N, Grossel P, Depasse F, Bainier C, Charraut D: DC scanning thermal microscopy: characterisation and interpretation of the measurement, *Int J Thermal Sci* 40(11):949–958, 2001.

Gourianova S, Willenbacher N, Kutschera M: Chemical force microscopy study of adhesive properties of polypropylene films: influence of surface polarity and medium, *Langmuir* 21(12):5429–5438, 2005.

Gruverman A, Tokumoto H, Prakash AS, Aggarwal S, Yang B, Wuttig M, Ramesh R, Auciello O, Venkatesan T: Nanoscale imaging of domain dynamics and retention in ferroelectric thin films, *Appl Phys Lett* 71(24):3492–3494, 1997.

Hafner JH, Cheung CL, Woolley AT, Lieber CM: Structural and functional imaging with carbon nanotube AFM probes, *Prog Biophys Mol Biol* 77(1):73–110, 2001.

Heaton MG, Prater CB, Kjaller KJ: *Lateral and chemical force microscopy mapping surface friction and adhesion*, disponível em: <http://www.bruker.jp/axs/nano/imgs/pdf/AN005.pdf>.

Hirose S, Nishita K, Niimi H: Influence of distribution of additives on electrical potential barrier at grain boundaries in ZnO-based multilayered chip varistor, *J Appl Phys* 100(8):083706/1–183706/, 2006.

Huey BD, Bonnell DA: Nanoscale variation in electric potential at oxide bicrystal and polycrystal interfaces, *Solid State Ionics* 131(1–2):51–60, 2000.

Image Metrology: *SPIP—scanning probe image processor*, disponível em: <http://www.imagemet.com/index.php?main=download>.

Jacobs HO, Knapp HF, Muller S, Stemmer A: Surface potential mapping: a qualitative material contrast in SPM, *Ultramicroscopy* 69(1):39–49, 1997.

Jak MJJ, Konstapel C, Van Kreuningen A, Verhoeven J, Van Gastel R, Frenken JWM: Automated detection of particles, clusters and islands in scanning probe microscopy images, *Surf Sci* 494(2):43–52, 2001.

Jalili N, Laxminarayana K: A review of atomic force microscopy imaging systems: application to molecular metrology and biological sciences, *Mechatronics* 14(8):907–945, 2004.

Jiang CS, Moutinho HR, Romero MJ, Al-Jassim MM, Kazmerski LL: Electrical charge trapping at defects on the Si(111)7×7 surface, *Appl Phys Lett* 88(6):022112/1–122112/, 2006.

Kalinin SV: *Nanoscale electric phenomena at oxide surfaces and interfaces by scanning probe microscopy*, Dissertation (Doctor of Philosophy), University of Pennsylvania, Pennsylvania, 2002, 298 pp.

Kalinin SV, Bonnell DA: Scanning impedance microscopy of electroactive interfaces, *Appl Phys Lett* 78(9):1306–1308, 2001.

Kalinin SV, Bonnell DA: Imaging mechanism of piezoresponse force microscopy of ferroelectric surfaces, *Phys Rev B* 65(12):125408/1–1125408/, 2002a.

Kalinin SV, Bonnell DA: Scanning impedance microscopy of an active Schottky barrier diode, *J Appl Phys* 91(2):832–839, 2002b.

Kalinin SV, Shao R, Bonnell DA: Local phenomena in oxides by advanced scanning probe microscopy, *J Am Ceram Soc* 88(5):1077–1098, 2005.

Kalinin SV, Rodriguez BJ, Jesse S, Shin J, Baddorf AP, Gupta P, Jain H, Williams DB, Gruverman A: Vector piezoresponse force microscopy, *Microsc Microanal* 12:206–220, 2006.

Kalyan Phani M, Kumar A, Arnold W, Samwer K: Elastic stiffness and damping measurements in titanium alloys using atomic force acoustic microscopy, *J Alloys Compd* 676:397–406, 2016.

Kämmer E, Götz I, Bocklitz T, Stöckel S, Dellith A, Cialla-May D, Weber K, Zell R, Dellith J, Deckert V, Popp J: Single particle analysis of herpes simplex virus: comparing the dimensions of one and the same virions via atomic force and scanning electron microscopy, *Anal Bioanal Chem* 408(15):4035–4041, 2016.

Kholkin A, Kalinin S, Roelofs A, Gruverman A: Review of ferroelectric domain imaging by piezoresponse force microscopy, Kalinin S, Gruverman A, editors: *Scanning probe microscopy: electrical and electromechanical phenomena at the nanoscale*, 1, New York, 2007a, Springer Science + Business Media, pp 173–214.

Kholkin AL, Bdikin IK, Shvartsman VV, Pertsev NA: Anomalous polarization inversion in ferroelectrics via scanning force microscopy, *Nanotechnology* 18(9):095502/1–195502/, 2007b.

Kondo Y, Osaka M, Benten H, Ohkita H, Ito S: Electron transport nanostructures of conjugated polymer films visualized by conductive atomic force microscopy, *ACS Macro Lett* 4(9):879–885, 2015.

Kopanski JJ, Marchiando JF, Rennex BG: Carrier concentration dependence of the scanning capacitance microscopy signal in the vicinity of p–n junctions, *J Vac Sci Technol B* 18(1):409–413, 2000.

Landau SA, Junghans N, Weiss PA, Kolbesen BO, Olbrich A, Schindler G, Hartner W, Hintermaier F, Dehm C, Mazure C: Scanning probe microscopy—a tool for the investigation of high-*k* materials, *Appl Surf Sci* 157(4):387–392, 2000.

Lee S: *Chemical functionalization of AFM cantilevers*, Dissertation, Chemical Engineering Seoul National University, 2002.

Lee JC, Chuang JH: A novel application of c-AFM: deep sub-micron single probing for IC failure analysis, *Microelectron Reliability* 43(9–11):1687–1692, 2003.

Leite FL, Mattoso LHC, Oliveira ON Jr, Herrmann PSP, Jr: The atomic force spectroscopy as a tool to investigate surface forces: basic principles and applications. In Méndez-Vilas A, Díaz J, editors: *Modern research, educational topics in microscopy*, Spain, 2007, Formatex, pp 747–757.

Leite FL, Hausen M, Oliveira GS, Brum DG, Oliveira ON: Nanoneurobiophysics: new challenges for diagnosis and therapy of neurologic disorders, *Nanomedicine* 10(23):3417–3419, 2015.

Lohr D, Bash R, Wang H, Yodh J, Lindsay S: Using atomic force microscopy to study chromatin structure and nucleosome remodeling, *Methods* 41(3):333–341, 2007.

Lu RP, Kavanagh KL, Dixon-Warren SJ, Springthorpe AJ, Streater R, Calder I: Scanning spreading resistance microscopy current transport studies on doped III–V semiconductors, *J Vac Sci Technol B* 20(4):1682–1689, 2002.

Marques VPB, Cilense M, Bueno PR, Orlandi MO, Varela JA, Longo E: Qualitative evaluation of active potential barriers in SnO_2-based polycrystalline devices by electrostatic force microscopy, *Appl Phys A* 87(4):793–796, 2007.

Martin Y, Williams CC, Wickramasinghe HK: Atomic force microscope force mapping and profiling on a sub 100-A scale, *J Appl Phys* 61(10):4723–4729, 1987.

Meyer F: Topographic distance and watershed lines, *Signal Process* 38:113–125, 1994.

Meyer E, Hug H, Bennewitz R: *Scanning probe microscopy: the lab on a tip*, Berlin, 2003, Springer.

Mironov VL: *Fundamentals of the scanning probe microscopy*, Nizhniy Novgorod, 2004, the Russian Academy of Sciences Institute of Physics of Microstructures.

Moraes AS, Pereira BBS, Moreau ALD, Hausen M, Garcia PS, Da Roz AL, Leite FL: Evidências de detecção do herbicida atrazina por espectroscopia de força atômica: uma ferramenta promissora para sensoriamento ambiental, *Acta Microscopica* 24(1):53–63, 2015.

Müller T: *Scanning tunneling microscopy: a tool for studying self-assembly and model systems for molecular devices*, disponível em: <http://www.veeco.com/pdfs/appnotes/AN85-STM_02085_rf_268.pdf>.

Nanotec Electronica: *WSxM. Version 5.0*, disponível em: <http://www.nanotec.es/products/wsxm/download.php>.

Nonnenmacher M, Oboyle MP, Wickramasinghe HK: Kelvin probe force microscopy, *Appl Phys Lett* 58(25):2921–2923, 1991.

Nonnenmacher M, Oboyle M, Wickramasinghe HK: Surface investigations with a Kelvin probe force microscope, *Ultramicroscopy* 42:268–273, 1992.

O'boyle MP, Hwang TT, Wickramasinghe HK: Atomic force microscopy of work functions on the nanometer scale, *Appl Phys Lett* 74(18):2641–2642, 1999.

Okada Y, Miyagi M, Akahane K, Kawabe M, Shigekawa H: Self-organized InGaAs quantum dots grown on GaAs (311)B substrate studied by conductive atomic force microscope technique, *J Cryst Growth* 245(3–4):212–218, 2002.

Oliveira GS, Leite FL, Amarante AM, Franca EF, Cunha RA, Briggs JM, Freitas LCG: Molecular modeling of enzyme attachment on AFM probes, *J Mol Graph Model* 45:128–136, 2013.

Pingree LSC, Reid OG, Ginger DS: Electrical scanning probe microscopy on active organic electronic devices, *Adv Mater* 21(1):19–28, 2009.

Prater CB, Maivald PG, Kjoller KJ, Heaton MG: *TappingMode imaging applications and technology*, disponível em: <http://www.tu-chemnitz.de/physik/OSMP/Soft/ws0506_ue03b.pdf>.

Rafael CG, Richard EW: *Digital image processing*, Upper Saddle River, 2002, Prentice-Hall.

Saurenbach F, Terris BD: Imaging of ferroelectric domain-walls by force microscopy, *Appl Phys Lett* 56(17):1703–1705, 1990.

Schonenberger C, Alvarado SF: Observation of single charge-carriers by force microscopy, *Phys Rev Lett* 65(25):3162–3164, 1990.

Schönherr H, Vancso GJ: Atomic force microscopy in practice. *Scanning force microscopy of polymers*, Berlin, 2010, Springer, pp 25–75.

Serry FM, Kjoller K, Thornton JT, Tench RJ, Cook D: *Electric force microscopy, surface potential imaging, and surface electric modification with the atomic force microscope (AFM)*, disponível em: <http://www.veeco.com/pdfs/appnotes/AN27_EFMSurfPot_260.pdf>.

Shao R, Kalinin SV, Bonnell DA: Local impedance imaging and spectroscopy of polycrystalline ZnO using contact atomic force microscopy, *Appl Phys Lett* 82(12):1869–1871, 2003.

Shin J, Meunier V, Baddorf AP, Kalinin SV: Nonlinear transport imaging by scanning impedance microscopy, *Appl Phys Lett* 85(18):4240–4242, 2004.

Smith DA, Connell SD, Robinson C, Kirkham J: Chemical force microscopy: applications in surface characterisation of natural hydroxyapatite, *Anal Chim Acta* 479(1):39–57, 2003.

Stadelmann T: *Review of scanning probe microscopy techniques*, 2016, disponível em: <http://timstadelmann.de/spmreview.pdf>.

Steffens C, Leite FL, Bueno CC, Manzoli A, Herrmann PSD: Atomic force microscopy as a tool applied to nano/biosensors, *Sensors* 12(6):8278–8300, 2012.

Steffens C, Leite FL, Manzoli A, Sandoval RD, Fatibello O, Herrmann PSP: Microcantilever sensors coated with a sensitive polyaniline layer for detecting volatile organic compounds, *J Nanosci Nanotechnol* 14(9):6718–6722, 2014a.

Steffens C, Manzoli A, Leite FL, Fatibello O, Herrmann PSP: Atomic force microscope microcantilevers used as sensors for monitoring humidity, *Microelectron Eng* 113:80–85, 2014b.

Steffens C, Manzoli A, Oliveira JE, Leite FL, Correa DS, Herrmann PSP: Bio-inspired sensor for insect pheromone analysis based on polyaniline functionalized AFM cantilever sensor, *Sens Actuators B Chem* 191:643–649, 2014c.

Stern JE, Terris BD, Mamin HJ, Rugar D: Deposition and imaging of localized charge on insulator surfaces using a force microscope, *Appl Phys Lett* 53(26): 2717–2719, 1988.

Suchodolskis A, Hallen A, Gran J, Hansen TE, Karlsson UO: Scanning spreading resistance microscopy of shallow doping profiles in silicon, *Nucl Instrum Methods Phys Res Sect B* 253(1–2):141–144, 2006.

Szunerits S, Pust SE, Wittstock G: Multidimensional electrochemical imaging in materials science, *Anal Bioanal Chem* 389(4):1103–1120, 2007.

Szymanski C, Wu CF, Hooper J, Salazar MA, Perdomo A, Dukes A, Mcneill J: Single molecule nanoparticles of the conjugated polymer MEH-PPV, preparation and characterization by near-field scanning optical microscopy, *J Phys Chem B* 109(18):8543–8546, 2005.

Terris BD, Stern JE, Rugar D, Mamin HJ: Contact electrification using force microscopy, *Phys Rev Lett* 63(24):2669–2672, 1989.

Trenkler T, De Wolf P, Vandervorst W, Hellemans L: Nanopotentiometry: local potential measurements in complementary metal–oxide–semiconductor transistors using atomic force microscopy, *J Vac Sci Technol B* 16(1):367–372, 1998.

Vasconcelos JS, Vasconcelos N, Orlandi MO, Bueno PR, Varela JA, Longo E, Barrado CM, Leite ER: Electrostatic force microscopy as a tool to estimate the number of active potential barriers in dense non-ohmic polycrystalline SnO_2 devices, *Appl Phys Lett* 89(15):152102/1–1152102/, 2006.

Wawkuschewski A, Cramer K, Cantow HJ, Magonov SN: Optimization of experiment in scanning force microscopy of polymers, *Ultramicroscopy* 58(2):185–196, 1995.

Webb HK, Truong VK, Hasan J, Crawford RJ, Ivanoya EP: Physico-mechanical characterisation of cells using atomic force microscopy—current research and methodologies, *J Microbiol Methods* 86(2):131–139, 2011.

Wittborn J, Rao KV, Nogues J, Schuller IK: Magnetic domain and domain-wall imaging of submicron Co dots by probing the magnetostrictive response using atomic force microscopy, *Appl Phys Lett* 76(20):2931–2933, 2000.

Xu Z, Hantschel T, Tsigkourakos M, Vandervorst W: Scanning spreading resistance microscopy for electrical characterization of diamond interfacial layers, *Phys Status Solidi A* 212(11):2578–2582, 2015.

Yalcin SE, Galande C, Kappera R, Yamaguchi H, Martinez U, Velizhanin KA, Doorn SK, Dattelbaum AM, Chhowalla M, Ajayan PM, Gupta G, Mohite AD: Direct imaging of charge transport in progressively reduced graphene oxide using electrostatic force microscopy, *ACS Nano* 9(3):2981–2988, 2015.

Zavala G: Atomic force microscopy, a tool for characterization, synthesis and chemical processes, *Colloid Polym Sci* 286:85–95, 2008.

Zavyalov VV, Mcmurray JS, Williams CC: Advances in experimental technique for quantitative two-dimensional dopant profiling by scanning capacitance microscopy, *Rev Sci Instrum* 70(1):158–164, 1999.

3

Spectroscopic Techniques for Characterization of Nanomaterials

Priscila Alessio, Pedro H.B. Aoki, Leonardo N. Furini, Alvaro E. Aliaga,
Carlos J. Leopoldo Constantino

STATE UNIVERSITY OF SÃO PAULO, SÃO PAULO, BRAZIL

CHAPTER OUTLINE

1 Ultraviolet–Visible Absorption

Ultraviolet–visible (UV–vis) absorption spectroscopy is a versatile technique that can be used for both the characterization and quantification of different types of organic, inorganic, and biological materials. The studies by Skoog et al. (2007) and Petrozzi and Goudie (2013) demonstrate one conceptual approach to the use of the UV–vis absorption technique. In particular, UV–vis is frequently applied to studies on the syntheses of new materials and nanoparticles in the field of nanotechnology through the use of colorimetric sensors in catalysis studies. Furthermore, the technique is also widely used for monitoring the growth of nanostructured thin films grown via various techniques such as Langmuir–Blodgett (LB) (Correia et al., 2012), layer-by-layer (LbL) (de Oliveira et al., 2011), Langmuir–Schaefer (LS) (Heriot et al., 2006), spray-LbL (Aoki et al., 2012), and vacuum thermal evaporation [physical vapor deposition (PVD)] (Volpati et al., 2008a, 2008b) among others.

1.1 Characterization of Materials and Nanoparticles

The synthesis of new materials, mainly carbon-based materials, has become a research field of great interest with the aim of creating materials with new properties and applications. Carbon nanotubes and graphene are two of the most thoroughly investigated materials in recent years, and they have been investigated for their mechanical, thermal, and electrical properties (Kuila et al., 2012; Scarselli et al., 2012). Fig. 3.1 shows the UV–vis absorption spectrum of multiwalled carbon nanotubes (MWCNTs) dispersed in water (Cheng et al., 2011) with absorption band maxima in the range of 240–260 nm (Fig. 3.1A) for MWCNTs with different lengths (l, large; m, medium; s, small). Fig. 3.1B shows the changes in the UV–vis absorption spectra of large MWCNTs in terms of different oxidation degrees, which are indicated by the nomenclature in the parentheses. Fig. 3.1C shows a deconvolution analysis of a UV–vis absorption spectrum. The analysis resolves the spectrum into three major bands, and the second band is the main band in both intensity and width. By analyzing the position and intensity of this band for different MWCNT samples, the authors were able to conclude that its intensity is related to variations in the oxidation degree and lengths of the MWCNTs. Based on resolved spectra, it is possible to identify and compare MWCNT samples from different sources (Cheng et al., 2011).

Another investigated field in nanotechnology is the synthesis, characterization, and application of colloidal nanoparticles, for example, organic materials, including polymers

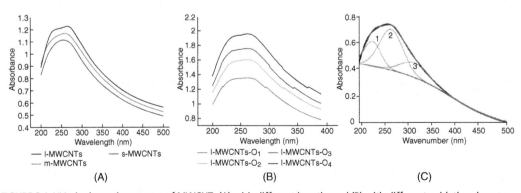

FIGURE 3.1 UV–vis absorption spectra of MWCNTs (A) with different lengths and (B) with different oxidation degrees. (C) A UV–vis absorption spectrum of MWCNTs resolved into three bands. *MWCNT*, Multiwalled carbon nanotube; *UV–vis*, ultraviolet–visible. *Adapted from Cheng X, et al: Characterization of multiwalled carbon nanotubes dispersing in water and association with biological effects, J Nanomater 2011, 2011.*

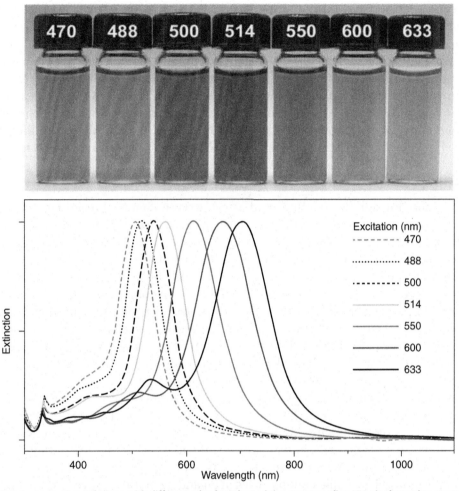

FIGURE 3.2 Ag nanoprism solutions with different edge lengths and the corresponding UV–vis absorption spectra. The numbers on the flask and spectra labels correspond to the irradiation wavelength used to prepare the nanostructures. *UV–vis*, Ultraviolet–visible. *Adapted from Xue C, Mirkin CA: pH-switchable silver nanoprism growth pathways,* Angew Chem Int Ed *46(12):2036–2038, 2007.*

(Yang et al., 2007), and ceramics with magnetic characteristics (Umut et al., 2012), especially metallic nanoparticles (Liu et al., 2011). UV–vis absorption is the main spectroscopy tool used to study these materials. A photochemical synthesis route that can be used to control the growth of Ag nanoprisms was reported by Xue and Mirkin (2007). Their methodology consists of adjusting the pH of the solution and exciting it at a specific wavelength, which allows for the synthesis of nanoprisms with uniform and controllable bands from the visible to the near-infrared absorption range. Fig. 3.2 shows a picture of the resulting nanoprisms and the corresponding extinction spectrum of the nanoprism obtained at pH 11.2 with an excitation wavelength between 488 and 633 nm.

The localized surface plasmon resonance (LSPR) properties of metallic nanoparticles influence their applications in catalysis studies, surface-enhanced Raman scattering (SERS), biosensors, and energy converters (Jans and Huo, 2012; Linic et al., 2011). The LSPR is characteristic of each metal because it depends on the resonance relationship with the incident radiation. An illustration of this effect can be seen in the UV–vis absorption spectra of the Au, Ag, and Cu spherical nanostructures shown in Fig. 3.3A (Linic et al., 2011). The LSPR spectrum is strongly dependent on the size and shape of the nanostructure in addition to the dielectric constant of the medium. Fig. 3.3B and C illustrates the dependence of the UV–vis absorption on the shape and size of the metallic nanostructures.

1.2 Catalysis

Nanostructured materials are often used as catalysts in chemical reactions. Changes in the shape and size of the catalytic materials can influence the catalytic activity of the nanostructured materials, and these changes can be measured using UV–vis absorption analyses. Zhang et al. (2010) studied the synthesis of microcrystals of Cu_2O as cubes, octahedra, and perfect 26- and 18-faced polyhedra. The photodegradation of methyl orange was investigated as a function of the shapes of the obtained structures using UV–vis absorption. The results showed that the 26- and 18-faced polyhedra had higher adsorption and photocatalytic activity than the octahedra and cube shapes (Fig. 3.4). The authors concluded that the performance of the catalysts can be increased by controlling the shape of the structures because the shape determines the number of atoms located at the edges, corners, or surfaces of the catalyst (Zhang et al., 2010).

Nanoparticles of bimetallic alloys have also been used as catalysts because they exhibit optical and catalytic properties that are dependent on their composition and morphology (He et al., 2010). Nanoparticles of bimetallic alloys can be used as mimetic enzymes in immunoassays, biocatalysis, and environmental monitoring because of their low cost, adjustable composition and structure, and high stability. Fig. 3.5 shows the influence of the composition of bimetallic nanoparticles of Pd and Ag on the catalysis of o-phenylenediamine dihydrochloride (OPD) tetramethylbenzidine (TMB), and 2,2′-azino-bis(3-ethylbenzothiazoline-6-sulfonic acid) diammonium salt (ABTS) (He et al., 2010).

1.3 Sensors

Functionalized nanostructures have been used as colorimetric sensors with UV–vis absorption. For example, Au nanoparticles (AuNPs) functionalized with glutathione have been used in the development of a portable Pb^{2+} sensor (Chai et al., 2010). The method allows for rapid, real-time detection of Pb^{2+}, and the experimental results showed that Pb^{2+} can be detected with a high sensitivity (100 nM) and selectivity over other heavy metal ions (Hg^{2+}, Mg^{2+}, Zn^{2+}, Ni^{2+}, Cu^{2+}, Co^{2+}, Ca^{2+}, Mn^{2+}, Fe^{2+}, Cd^{2+}, Ba^{2+}, and Cr^{3+}) (Chai et al., 2010).

Another application of the UV–vis absorption technique for nanostructure sensors was presented by Li et al. (2010a, 2010b) for the detection of nucleotide polymorphisms. A DNA replication technique called isothermal rolling circle amplification (RCA), which is

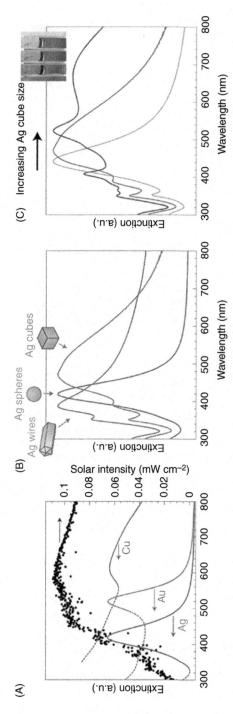

FIGURE 3.3 (A) Normalized extinction spectra of spherical metal nanoparticles: Ag (38 ± 12 nm in diameter), Au (25 ± 5 nm); and Cu (133 ± 23 nm); (B) normalized extinction spectra for Ag nanoparticles in cube, wire, and sphere shapes; (C) normalized extinction spectra for Ag nanocubes as a function of size. The inset shows a photograph of three nanocube samples suspended in ethanol. *Adapted from Linic S, Christopher P, Ingram DB: Plasmonic-metal nanostructures for efficient conversion of solar to chemical energy,* Nat Mater *10(12):911–921, 2011.*

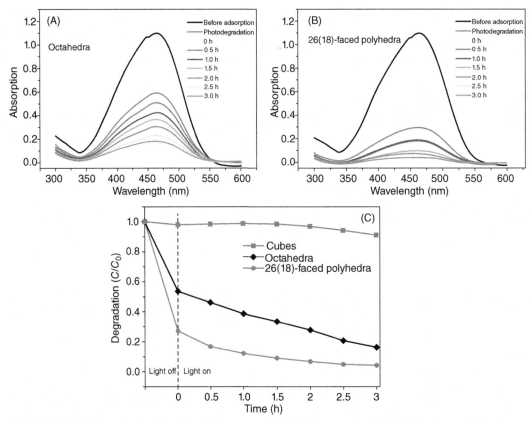

FIGURE 3.4 UV–vis absorption spectra of a methyl orange solution in the presence of Cu$_2$O in the shape of (A) octahedra and (B) 26- and 18-faced polyhedra. (C) Photodegradation curves based on the UV–vis absorption intensity of methyl orange in the presence of Cu$_2$O as polyhedra, cubes, and octahedra. *UV–vis*, Ultraviolet–visible. *Adapted from Zhang Y, et al: Shape effects of Cu2O polyhedral microcrystals on photocatalytic activity,* J Phys Chem *114(11):5073–5079, 2010.*

used for the highly sensitive detection of nucleic acids and proteins, was combined with a colorimetric method based on AuNPs to achieve detection limits on the order of femtomoles (Li et al., 2010a, 2010b).

AuNPs were also used in the colorimetric detection of melamine in milk (Li et al., 2010a, 2010b). The detection mechanism is based on the ability of melamine to induce aggregation in AuNPs, which results in a shift of the absorption band (LSPR) from blue to red, as shown in Fig. 3.6A. The detection limit was 0.4 mg/L.

1.4 Monitoring the Growth of Nanostructured Films

The growth of nanostructured thin films can be monitored using UV–vis and the Beer–Lambert law, which states that the UV–vis absorbance is directly proportional to the concentration of a substance in a medium: $A = \alpha \times l \times c$, where A is the absorbance, α is the

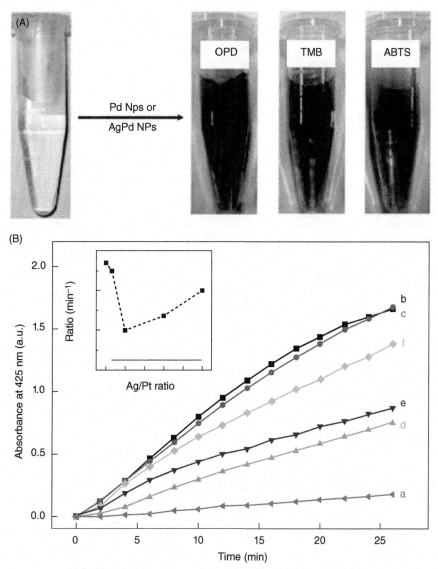

FIGURE 3.5 (A) Evolution of colors in the oxidation of OPD, TMB, and ABST catalyzed by Pd (PdNC) or AgPd (AgPdNC) nanocrystals. (B) Absorbance at 425 nm as a function of time after addition of the same PdNC (curve b) and AgPdNC aliquots with Ag:Pd ratios of 1:3 (curve c), 1:1 (curve d), 3:1 (curve e), and 5:1 (curve f). Curve "a" represents a control experiment without a catalyst. The inset in "B" shows the dependence of the OPD oxidation rate on the Ag:Pd ratio (dashed line), and the solid line indicates the control experiment value. Reaction conditions: 0.3 M H_2O_2, 0.3 mM OPD, and 26 μM of PdNC or AgPdNC at 40°C. *OPD*, o-Phenylenediamine dihydrochloride; *TMB*, tetramethylbenzidine. *Adapted from He W, et al: Design of AgM bimetallic alloy nanostructures (M = Au, Pd, Pt) with tunable morphology and peroxidase-like activity, Chem Mater 22(9):2988–2994, 2010.*

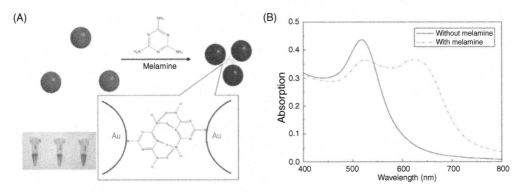

FIGURE 3.6 (A) Representation of the colorimetric mechanism for melamine detection. The inset shows the photos for the solutions: (1) 400 μL Au nanoparticles + 20 μL H_2O, (2) 400 μL Au nanoparticles + 20 μL melamine (5×10^{-3} g/L), and (3) 400 μL Au nanoparticles + 20 μL melamine (20×10^{-3} g/L). Experimental conditions: 1.4 μM Au nanoparticles, incubation time of 1 min, and room temperature (~20°C). (b) UV–vis absorption spectra of the Au nanoparticles in the absence (solid line) and presence (dashed line) of melamine. *UV–vis*, Ultraviolet–visible. *Adapted from Li J, et al: Rolling circle amplification combined with gold nanoparticle aggregates for highly sensitive identification of single-nucleotide polymorphisms, Anal Chem 82(7):2811–2816, 2010; Li L, Li B, Cheng D, Mao L: Visual detection of melamine in raw milk using gold nanoparticles as colorimetric probe, Food Chem 122(3):895–900, 2010.*

absorption coefficient characteristic of each substance, l is the distance the light travels (optical path), and c is the analyte concentration (Skoog et al., 2007). To monitor the growth of nanostructured thin films, a UV–vis absorption spectrum is taken for each layer (LB films) or bilayer (LbL films) or for a given deposited mass (e.g., an evaporated film). The relationship between the absorbance and the number of layers (or bilayers or mass) may or may not be linear. A linear relationship indicates that the same amount of material is deposited on the solid substrate during the deposition of each layer or bilayer. Examples of linear growth can be seen in Fig. 3.7A and B (Volpati et al., 2008a, 2008b) for a PVD film and an LB film of iron phthalocyanine (FePc), respectively. However, the UV–vis absorption spectra of FePc are different, which indicates that different nanostructuring (supramolecular organization) is obtained from the different preparation techniques used to create these thin films.

A nonlinear relationship between the absorbance and the number of layers (or bilayers or mass) indicates that different adsorption processes are controlling the film growth. For example, Fig. 3.8A (de Oliveira et al., 2011) shows the growth analysis of an LbL film composed of two enzymes [glucose oxidase (GOx) and invertase (INV)] and a polyelectrolyte [poly(allylamine hydrochloride) (PAH)] in a tetralayer structure (PAH/GOx/PAH/INV). The growth behavior of this film is the sum of two exponential functions, which suggests that two processes with different characteristic times are involved in the film adsorption. The authors compared the growth of this bienzymatic film to the growth of monoenzymatic films (PAH/GOx) and (PAH/INV), and all the films showed linear growth. The exponential growth of the tetralayer film is associated with the presence of both enzymes in the film.

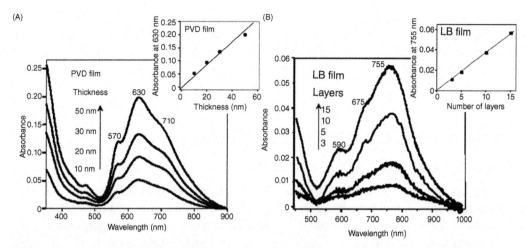

FIGURE 3.7 UV–vis absorption spectra for (A) FePc PVD films with different mass thicknesses and (B) FePc LB films with different numbers of layers. The insets show (A) absorbance at 630 nm versus the mass thickness for the FePc PVD film and (B) absorbance at 755 nm versus the number of FePc LB film layers. *FePC*, Iron phthalocyanine; *LB*, Langmuir–Blodgett; *PVD*, physical vapor deposition; *UV–vis*, ultraviolet–visible. *Adapted from Volpati D, et al: Exploiting distinct molecular architectures of ultrathin films made with iron phthalocyanine for sensing,* J Phys Chem B *112(48):15275–15282, 2008; Volpati D, Job AE, Aroca RF, Constantino CJL: Molecular and morphological characterization of bis benzimidazo perylene films and surface-enhanced phenomena,* J Phys Chem B *112(13):3894–3902, 2008.*

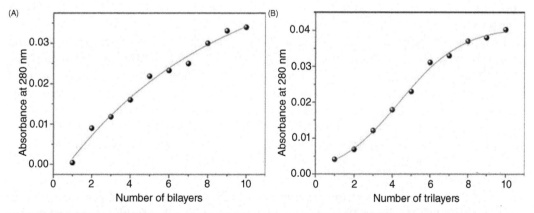

FIGURE 3.8 (A) Absorbance at 280 nm for the LbL film in tetralayers of PAH/GOx/PAH/INV versus the number of deposited layers; (B) absorbance at 280 nm for the LbL film in trilayers of PAH/GOx/INV versus the number of deposited layers. *GOx*, Glucose oxidase; *INV*, invertase; *LbL*, layer-by-layer; *PAH*, poly(allylamine hydrochloride). *Adapted from de Oliveira RF, De Moraes ML, Oliveira ON, Ferreira M: Exploiting cascade reactions in bienzyme layer-by-layer films,* J Phys Chem C *115(39):19136–19140, 2011.*

In the case of the bienzymatic film in the trilayer structure (PAH/GOx/INV) shown in Fig. 3.8B, the relationship between the absorbance and the number of trilayers showed Boltzmann sigmoidal behavior, which is not typical of LbL films. This may be caused by the rearrangement of biomolecules in layers after a certain number of layers are deposited (de Oliveira et al., 2011).

2 Fourier Transform Infrared Spectroscopy

2.1 Molecular Interactions

In recent decades, one of the most promising fields in materials science that has been strongly developed is technology for the manipulation of materials on a nanometer scale. This technology has made it possible to revisit materials with interesting macroscopic characteristics to examine their nanometer-scale properties (Centurion et al., 2012). For thin films, manipulation on the nanometer scale can be achieved via immobilization of different materials on the films using different techniques (Blodgett, 1935; Decher, 1997; Langmuir, 1917; Volpati et al., 2008a, 2008b). Fourier transform infrared (FTIR) spectroscopy is widely used for the identification of materials in films and provides information on the interaction mechanisms of analytes of interest and molecular organization (Aoki et al., 2012; Pavinatto et al., 2011; Volpati et al., 2008a, 2008b).

For example, phospholipids are used as simple mimetic systems to study the cellular membrane (Aoki et al., 2009a, 2009b). Langmuir and LB films are among the main techniques used for the production of phospholipid ordered layers structured in monolayers or bilayers in aqueous subphases and solid substrates. However, the difficulties encountered in the production of LB films containing multilayers of phospholipids limit the application of this technique, and the technique depends on the sensitivity of the experimental analysis to be performed. Thus, the development of methodologies for the deposition of multilayered phospholipids is desired. Aoki et al. (2009a, 2009b) demonstrated the deposition of LB multilayers of the phospholipid dipalmitoylphosphatidylglycerol (DPPG) by adding the polyelectrolyte PAH to the aqueous subphase. The surface pressure isotherms versus the mean molecular area demonstrated that the polyelectrolyte tends to interact with the phospholipid monolayer. The FTIR spectra were obtained for the LB films containing multilayers of (DPPG + PAH) and were compared with the FTIR spectra of cast films of DPPG and PAH, as shown in Fig. 3.9. The main differences in the LB film spectra were the displacements of the bands allocated to the PO_4^- group (1221, 1094, and 1048 cm^{-1}) of the DPPG and NH_3^+ group (1551 and 1607 cm^{-1}) of the PAH compared to the cast films. These differences strongly indicated the existence of electrostatic interactions between the DPPG (PO_4^-) and PAH (NH_3^+), which were responsible for the growth of the LB multilayers of (DPPG + PAH), as shown in detail in Fig. 3.9.

Centurion et al. (2012) reported the production of thin films with active elements for the development of humidity sensors. Nanostructured films of cobalt tetrasulfonated phthalocyanine (CoTsPc) were deposited in alternating layers with the polyelectrolytes

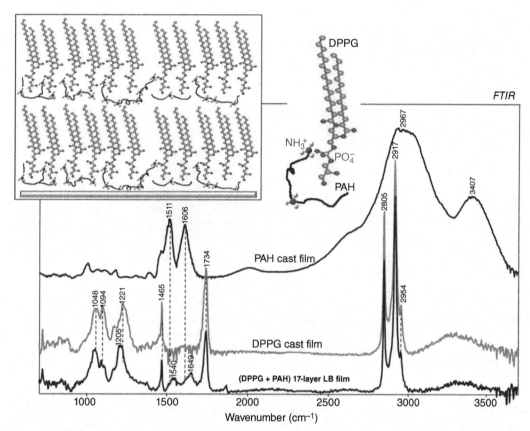

FIGURE 3.9 FTIR spectra of 17 LB layers of (DPPG + PAH) deposited on ZnSe. The FTIR spectra of the cast films of PAH and DPPG are given as reference. The inset shows the molecular architecture proposed for the LB films containing (DPPG + PAH). *DPPG*, Dipalmitoylphosphatidylglycerol; *FTIR*, Fourier transform infrared; *LB*, Langmuir–Blodgett; *LbL*, layer-by-layer; *PAH*, poly(allylamine hydrochloride). *Adapted from Aoki PHB, et al: Layer-by-layer technique as a new approach to produce nanostructured films containing phospholipids as transducers in sensing applications,* Langmuir *25(4):2331–2338, 2009; Aoki PHB, et al: Taking advantage of electrostatic interactions to grow Langmuir–Blodgett films containing multilayers of the phospholipid dipalmitoylphosphatidylglycerol,* Langmuir *25(22):13062–13070, 2009.*

poly(amidoamine) dendrimers (PAMAM) and PAH using the LbL technique (Decher, 1997). The FTIR measurements revealed the existence of electrostatic interactions between the CoTsPc sulfonic groups and the polyelectrolytes amine groups, and these interactions were essential for the growth of the films. Stronger interactions, such as covalent bonds, may also be responsible for the growth of the LbL films. In such cases, the changes in the FTIR spectra are more intense and may include the displacement and variation in the relative band intensity and the disappearance and/or appearance of new bands. In a study by Hu and Ji (2011), the LbL multilayer films were deposited via alternating covalent bonds with the aim of applying the films to the controlled release of therapeutic agents.

Crespilho et al. (2009) described the synthesis of biological hybrid materials with 3D nanostructures that were created using AuNPs and methionine (Met). Met is an essential

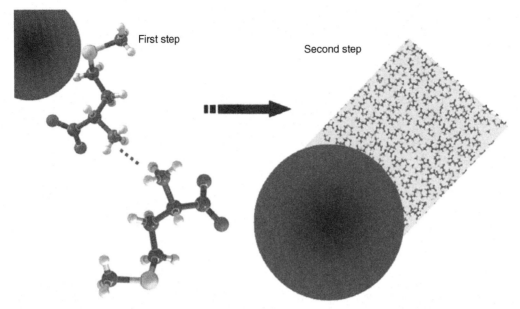

FIGURE 3.10 Schematic representation of the mechanisms of interaction between AuNPs and Met. *AuNP,* Au nanoparticle; *Met,* methionine. *Adapted from Crespilho FN, et al: The origin of the molecular interaction between amino acids and gold nanoparticles: a theoretical and experimental investigation,* Chem Phys Lett *469(1–3):186–190, 2009.*

amino acid in proteins and is responsible for molecular biosynthesis. The type of nanostructure formed can be controlled via the intermolecular interactions between Met and AuNPs, and these interactions strongly depend on the pH used during the synthesis. Understanding the mechanisms involved in the formation of these hybrid nanostructures is not easy because nonspecific (electrostatic) interactions, specific (H bond) interactions, and even molecular recognition processes all may occur. Thus, FTIR spectroscopy and computational simulations were used to determine the AuNP–Met interactions that allow the formation of hybrid nanostructures. The goal was to compare the Met FTIR spectra with the AuNP–Met nanostructure spectra to search for spectral changes that could indicate the possible interactions. The results show that in the first stage, the AuNPs and Met interact via sulfonic groups, as shown in Fig. 3.10. In the second stage, the 3D nanostructures are formed via the reorientation of the Met molecules so the amine groups interact through H bonds.

2.2 Molecular Orientation

Organic semiconductors have been successfully applied in optoelectronics and in electronic devices such as solar cells, field-effect transistors (FETs), and light-emitting diodes (LEDs) (Aljada et al., 2012; Cotrone et al., 2012; Zafer et al., 2007). The application of these materials in electronic devices can be accomplished using thin films. The molecular

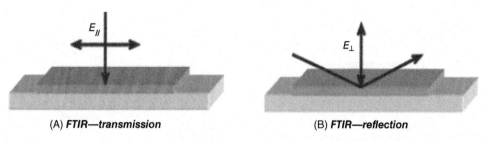

(A) **FTIR—transmission**   (B) **FTIR—reflection**

FIGURE 3.11 (A) Transmission mode: electric field ($E_{//}$) of the light parallel to the substrate surface; (B) reflection mode: the angle of light incidence (~80 degrees) maximizes the electric field component ($E_\perp$) perpendicular to the substrate surface. *FTIR*, Fourier transform infrared. *Adapted from Volpati D, et al: Exploiting distinct molecular architectures of ultrathin films made with iron phthalocyanine for sensing, J Phys Chem B 112(48):15275–15282, 2008; Volpati D, Job AE, Aroca RF, Constantino CJL: Molecular and morphological characterization of bis benzimidazo perylene films and surface-enhanced phenomena, J Phys Chem B 112(13):3894–3902, 2008.*

organization of the films plays an important role in the electrical and optical proper-ties and in the device performance. Vibrational spectroscopy, that is, FTIR and Raman scattering, is an important tool for the structural characterization of thin films (Del Cano et al., 2002; Kam et al., 2001).

Molecular organization can be determined from FTIR spectra obtained in reflec-tion and transmission mode when combined with the surface selection rules (Born and Wolf, 1987; Bradshaw and Schweizer, 1988; Debe, 1987). The rules are briefly described as follows: (1) In transmission mode, the electric field of the incident radiation is parallel to the substrate surface ($E_{//}$) because the propagation direction of the radiation beam is perpendicular to the substrate surface. (2) In reflection mode, considering the metal (Ag or Au) and the radiation incidence angle (−80 degrees), the electric field is prefer-ably polarized perpendicular to the substrate surface ($E_\perp$). The transmission and reflection modes are shown in Fig. 3.11A and B, respectively, and were adapted from Volpati et al. (2008a, 2008b). (3) The absorbed radiation intensity (*I*) is given by the scalar product be-tween the electric field (*E*) and the change in the molecular dipole moment (*μ*), that is, $I = \vec{E} \times \vec{u}$. Because the direction of the incident radiation electric field is known in both the transmission and reflection modes, the molecular organization can be determined using the FTIR bands whose *μ* components are well established. Thus, the *μ* component parallel to the substrate surface is stronger in the transmission mode ($E_{//}$ is parallel to *μ*), and the *μ* component perpendicular to the substrate surface is stronger in the reflection mode ($E_\perp$ is parallel to *μ*). (4) The spectrum of the powder material (KBr pellet) is provided as a refer-ence system for random molecular orientation.

The study by Volpati et al. (2008a, 2008b) is an example of molecular orientation de-termination for the organic semiconductor bisbenzimidazole perylene (AzoPTCD) thin films deposited using vacuum thermal evaporation (PVD). Fig. 3.12A shows the spectra obtained in transmission and reflection modes for a 40-nm-thick AzoPTCD PVD film. The FTIR spectrum obtained in the transmission mode for the AzoPTCD powder dispersed in a KBr pellet is shown as a reference. The differences between the relative intensities of the

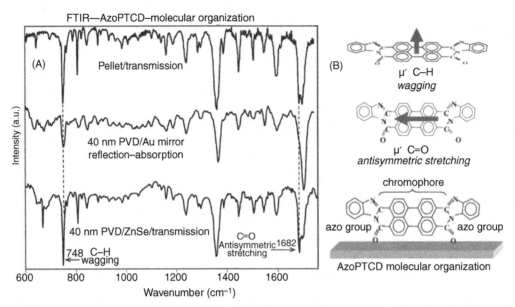

FIGURE 3.12 (A) FTIR spectra obtained in transmission mode for AzoPTCD powder dispersed in a KBr pellet for a 40-nm-thick PVD film deposited on ZnSe. FTIR spectrum obtained in reflection mode for a 40-nm-thick PVD film deposited on an Au mirror. (B) μ component of the C–H wagging vibration (out of the plane) and C=O stretching (in the plane), and the molecular organization determined for the AzoPTCD PVD films. *FTIR,* Fourier transform infrared; *PVD,* physical vapor deposition. *Adapted from Volpati D, et al: Exploiting distinct molecular architectures of ultrathin films made with iron phthalocyanine for sensing,* J Phys Chem B *112(48):15275–15282, 2008; Volpati D, Job AE, Aroca RF, Constantino CJL: Molecular and morphological characterization of bis benzimidazo perylene films and surface-enhanced phenomena,* J Phys Chem B *112(13):3894–3902, 2008.*

PVD film spectra (in transmission and reflection modes) suggest a strong anisotropy in these films in terms of the molecular organization. For example, the μ component of the bands at 748 and 1682 cm^{-1} (C–H wagging vibration and C=O asymmetrical stretching, respectively) is shown in Fig. 3.12. Because the relative intensity of the band at 748 cm^{-1} is stronger in the transmission mode (Fig. 3.12), it can be concluded that the AzoPTCD molecule is found with the chromophore plane perpendicular to the substrate surface. Moreover, the band at 1682 cm^{-1} has a much higher relative intensity in the transmission mode than in the reflection mode (Fig. 3.12). This confirms not only that the AzoPTCD preferably organizes in a perpendicular manner to the substrate but also that the molecule is supported by its major axis, as shown in Fig. 3.12.

The molecular organization is not governed only by the attractive and repulsive forces between the molecules. Experimental factors, such as the deposition rate of the films, the substrate temperature, and vacuum level, may be adjusted to induce different molecular organizations. External factors may also influence the molecular organization of films. Zanfolim et al. (2010) demonstrated the effects on the molecular organization of zinc phthalocyanine (ZnPc) PVD films when the films were heated to a temperature of 200°C under an ambient atmosphere for 2 h. Before heating, the ZnPc molecules preferably arranged

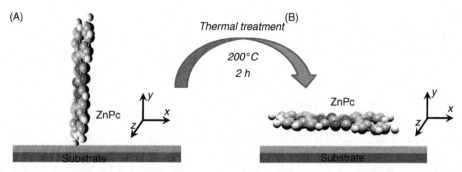

FIGURE 3.13 Molecular organization of ZnPc PVD films (A) before and (B) after heating at 200°C for 2 h. *PVD*, physical vapor deposition; *ZnPc*, zinc phthalocyanine.

in a macrocycle ring perpendicular to the substrate surface, as shown in Fig. 3.13A. After heating, a drastic change in the molecular organization of the film was observed. The ZnPc molecules assumed an organization with a macrocycle virtually parallel to the substrate surface, as shown in Fig. 3.13B.

Thin film manufacturing techniques are also important in molecular organization. Different techniques can induce different molecular organizations. Volpati et al. (2008a, 2008b) reported on thin films of FePc that were manufactured using the PVD, LB, and LbL techniques. Although the films were composed of the same material, the molecular organization in each film was different. The LbL film was isotropic, that is, it did not show any molecular organization. The PVD films had FePc molecules that were positioned between 45 and 90 degrees relative to the substrate. In the LB films, the FePc molecules were positioned between 0 and 45 degrees relative to the substrate.

This approach is used to determine the molecular organization in thin films of different classes of materials including small molecules, such as phthalocyanines (Alessio et al., 2010; Zanfolim et al., 2010) and perylenes (Del Cano et al., 2004), and macromolecules, such as lignins (Pasquini et al., 2002; Pereira et al., 2007) and luminescent polymers (Ferreira et al., 2003).

3 Raman Scattering

3.1 Carbon-Based Nanomaterials

Raman spectroscopy, which uses the inelastic scattering of the light falling on a material, is widely used for the characterization of materials in various fields because it is a nondestructive, fast, efficient, and easy technique. It has been widely applied to the characterization of nanostructures. Carbon-based materials, for example, can be characterized using Raman spectroscopy to indicate the formation of nanostructures. Hollow carbon nanopolyhedrons synthesized at low temperature (Zhu et al., 2012) are characterized by the presence of the D band at 1328 cm^{-1}, which is attributed to the presence of defects and disorder in the carbonaceous materials, and the G band at 1579 cm^{-1}, which is attributed to C–C stretching. One parameter used to evaluate the degree of disorder is the I_D/I_G ratio (where I_D and I_G are the intensities of the D and G bands, respectively) (Zhu et al., 2012). In the case of the nanopolyhedrons, the I_D/I_G ratio is 1.15, which indicates the absence of long-range order and confirms the formation of the nanostructures (Zhu et al., 2012).

Graphite with ferromagnetic properties was studied using Raman spectroscopy by Pardo et al. (2012). The authors proposed a method to modify the graphite to obtain stable ferromagnetic graphite at room temperature. The graphite was modified using controlled oxidation with copper oxide (CuO). The characterization using Raman spectroscopy with a laser source at 514.5 nm was performed on two types of graphite (modified and unmodified). In the modified graphite spectrum, a band appears at 3480 cm^{-1} and is attributed to the ferromagnetic property. The ferromagnetic character is obtained by introducing defects, such as pores and corners, which are revealed by the Raman spectrum. The increasing intensity of the D band suggests an increase in the disorder and a decrease in the grain size, which is supported by the comparison of the I_D/I_G ratio between the modified and unmodified graphite.

Raman spectroscopy can be applied to evaluate the electrical characteristics of a double layer on carbonaceous materials (Chang et al., 2012). The device (capacitor) is structured with a carbon sheet where the carbon nanowall (CNW) is deposited, which is followed by a Ni film deposition (Chang et al., 2012). The device is heated to 750°C to form nickel oxide (NiO) nanoparticles. Fig. 3.14A–C shows the scanning electron microscopy (SEM) images of the carbon sheet before and after the immobilization of the carbon nanostructures (CNW/carbon sheet), followed by NiO synthesis on the CNW (NiO/CNW/carbon

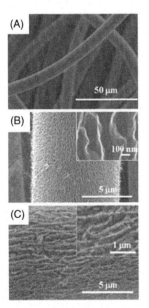
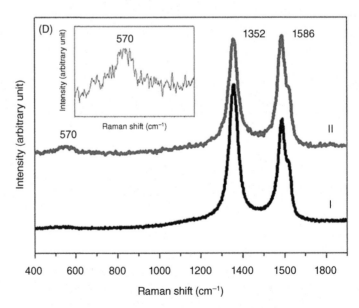

FIGURE 3.14 SEM images of the (A) carbon sheet, (B) CNW/carbon sheet, and (C) NiO/CNWs/carbon sheet. Raman spectra of the (D) CNW/carbon sheet (spectrum I) and NiO/CNWs/carbon sheet (spectrum II). *CNW*, carbon nanowall; *SEM*, scanning electron microscopy. *Adapted from Chang HC, et al: Preparation and electrochemical characterization of NiO nanostructure–carbon nanowall composites grown on carbon cloth,* Appl Surf Sci *258(22):8599–8602, 2012.*

sheet), and the insets show a zoom of the area. Fig. 3.14D shows the Raman spectra of the CNW/carbon sheet (spectrum I) and NiO/CNW/carbon sheet (spectrum II). The results show two major bands, the G band at 1586 cm^{-1}, which is attributed to the presence of graphite with sp^2-C bonds, and the D band at 1352 cm^{-1}, which is attributed to the system disorder. The CNW/carbon sheet has a band at 1352 cm^{-1} that is more intense than that at 1586 cm^{-1}. After Ni fixation and formation of the NiO nanoparticles, the band at 1352 cm^{-1} becomes less intense than the band at 1586 cm^{-1}. The presence of the band at 570 cm^{-1}, featured in Fig. 3.14D, is attributed to the presence of the NiO nanoparticles.

One of the most popular materials in the nanoscience field is carbon nanotubes because of their electrical and mechanical properties and chemical stability. Using Raman spectroscopy, it is possible to determine the nanotube diameters (Graupner, 2007), nanotube–nanotube interactions (Rao et al., 2001), and thermal conductivity. The analyses are usually performed considering the position, intensity, and full width at half-maximum of the peaks. Carbon nanotubes can be single-walled (SWCNTs) or multiwalled (MWCNTs). The determination of thermal conductivity is performed by analyzing the displacement of the G-band position, which is attributed to the thermal expansion and weakening of the C–C bond (Kim et al., 2011). Fig. 3.15 shows the G-band dependence on the temperature. The results indicate that the G band shifts to shorter wavelengths with increasing temperature. The SWCNTs produced using an arc discharge had the highest thermal conductivity values, 65.9 W/mK (Kim et al., 2011). In general, powder SWCNTs characterized using Raman spectroscopy have specific bands located at approximately 160 and 1590 cm^{-1} when irradiated with a laser at 1064 nm (Rao et al., 2001). The low-frequency bands are strongly dependent on the nanotube diameter.

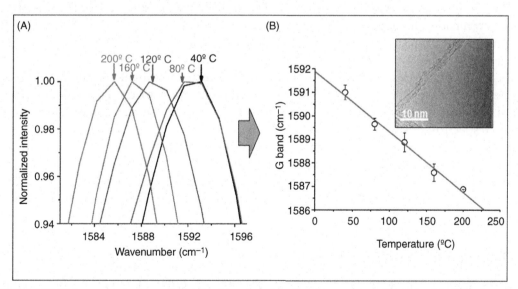

FIGURE 3.15 (A) Raman spectrum monitoring the G band as a function of temperature; (B) graph of the G band position as a function of temperature. The inset shows a transmission electron microscopy (*TEM*) image of a carbon nanotube (*SWCNT*). SWCNT, single-walled carbon nanotube. *Adapted from Kim D, et al: Raman characterization of thermal conduction in transparent carbon nanotube films,* Langmuir *27(23):14532–14538, 2011.*

It is also possible to characterize carbon nanotube structures or composites with hybrid characteristics, and Raman spectroscopy is useful to evaluate the possible chemical or physical interactions between the materials. For example, Bokobza and Zhang (2012) characterized pure MWCNTs and composites of MWCNTs/rubber. The effects of laser irradiation on the sample were also investigated. MWCNTs were purchased commercially and immobilized in an elastomer matrix (styrene-butadiene rubber). The Raman spectra were obtained with a 514.5-nm laser. As previously discussed, there are two major bands for carbon nanotubes, one located at 1580 cm^{-1} (G band), which is attributed to the vibration in the C–C bond plane with a shoulder at 1604 cm^{-1} (D′ band), and the other band at 1342 cm^{-1} (D band), which is attributed to the disorder of the structure. There is also the G′ band, which is attributed to the overtone of the D band. When assessing the effect of the interaction between the MWCNTs and comparing dispersed MWCNT spectra with nondispersed spectra, the latter are shifted to longer wavelengths, which indicate less interaction. The Raman spectrum of the MWCNTs/rubber composite is dominated by the MWCNTs. The bands pertaining to the matrix shifted to longer wavelengths because of the physical limitations of the polymer chains introduced by the MWCNTs. One characteristic is the linear dependence of the excitation energy source on the wave number of the D and G′ bands. As the excitation source changed from 785 to 514.5 and 457.9 nm, the wave number of the D band increased, but the I_D/I_G ratio decreased, which indicated a dispersive energy behavior. The effect of temperature was also evaluated by maintaining a laser with a wavelength of 514.5 nm and irradiating the MWCNTs with 1%, 10%, 25%, 50%, and 100% of the maximum power (10 mW). The authors noted that the band intensity depends on the power, and it was more intense as the power increased. For a greater power (higher temperature), there was a shift to shorter wavelengths for both bands (G and D). This process is reversible and does not cause damage to the MWCNTs.

In regards to physical interactions, Raman spectroscopy is sensitive to SWCNTs linked to Si or SiO_2 ordered nanopillars on Si substrates (Kobayashi et al., 2004). The Raman signals are the most intense when the SWCNTs are connected to the pillars, and new bands appear in low-frequency regions, indicating the connection to the pillars (Kobayashi et al., 2004). The functionalization of nanotubes via doping can also be detected using Raman spectroscopy. SWCNTs are p- or n-type semiconductors, depending on their functionalization. Suzuki and Hibino (2011) reported the characterization of SWCNTs doped with boron, which created p-type semiconductors and n-type nitrogen. Functionalization via acid treatment of MWCNTs can also be measured (Mantha et al., 2010) using the increasing I_D/I_G ratio, which increased from 1.36 (pure MWCNTs) to 1.7 [functionalized MWCNTs (f-MWCNTs)].

3.2 Nanomaterials From Metals

Raman spectroscopy can also be used to characterize different nanostructures made from the same compound, for example, zinc sulfide (ZnS) (Kim et al., 2012). Different ZnS nanostructures, such as nanowires, nanocombs, and nanobelts, were studied and characterized using Raman spectroscopy. The SEM images of the nanostructures are shown in Fig. 3.16A–C. The nanowires are shown in Fig. 3.16A, the nanocombs in Fig. 3.16B, and the nanobelts in Fig. 3.16C. The Raman spectra of the nanostructures are shown in Fig. 3.16D for the nanowires, in Fig. 3.16E for the nanocombs, and in Fig. 3.16F for the nanobelts. The three structures (nanowires, nanobelts, and nanocombs) were irradiated with two excitation source energies (1.96 and 2.71 eV, which refer to 632.8 and 457.9 nm, respectively). The Raman spectroscopy analysis indicated that the signal/noise ratio for the nanowires was lower when they were irradiated with 2.71 eV, and they also showed more bands between 200 and 300 cm^{-1}. The peak at 521 cm^{-1} is due to the Si substrate, and the disappearance of this peak in the other spectra is due to the dense growth of nanocombs and nanobelts. The wave number change from 350.2 to 349.2 cm^{-1} during the formation of the nanobelts indicates the generation of elastic tension.

The presence of defects on the surfaces of nanomaterials makes them capable of adsorbing O_2, and it is possible to monitor the presence of this gas using Raman spectroscopy. Wu et al. (2010) characterized nanocrystals of CeO_2 with different shapes: nanorods, nanocubes, and nanooctahedra. The transmission electron microscopy (TEM) images (high resolution) of the nanorods are shown in Fig. 3.17A and B, the nanocubes in Fig. 3.17C and D, and the nanooctahedra in Fig. 3.17E and F. In the Raman spectroscopy characterization, two sources of excitation were used, 532 and 325 nm. The results were better with excitation at 325 nm because of the resonance Raman effect. CeO_2 has a major band at 462 cm^{-1} and three less intense ones at 258, 595, and 1179 cm^{-1}. In the nanorods, there is band broadening at 462 cm^{-1} due to the size distribution of the nanostructures. The characterization of the surface defects was accomplished using the presence of adsorbed O_2. The results indicated that in the calcined (heated) samples, the Raman bands were not observed because of the presence of adsorbed O_2. This indicates that the adsorption of O_2 does not occur in the oxidized samples, and the samples subjected to reduction do adsorb O_2. Fig. 3.17G shows the Raman spectra for the reduced and heated nanorods at 673, 773, and 873 K. The Raman spectrum of the sample subjected to reduction after heating at 673 K has three bands, 1139, 862 (shoulder), and 830 cm^{-1}, which are attributed to the O_2 adsorbed on the surface. In the Raman spectrum of the sample reduced after heating at 773 K, the band at 1139 cm^{-1} is barely observed, and the shoulder at 862 cm^{-1} is more intense when compared to the major band at 830 cm^{-1}. The adsorbed O_2 bands were not observed in the samples treated at 873 K. When $^{18}O_2$ is used instead of $^{16}O_2$, the bands at 830 and 862 cm^{-1} shift to 782 and 813 cm^{-1}, as shown in Fig. 3.17G. Raman spectroscopy is a useful technique for characterizing surface defects on nanomaterials. Roro et al. (2012) reported MWCNTs that were chemically functionalized and mixed with a NiO solution. Raman spectra were obtained for the pure MWCNTs, f-MWCNTs, and nanocomposites with NiO (f-MWCNTs/NiO). The results showed that the D and G band peaks were broader and the I_D/I_G ratio was greater for the nanocomposites compared to those for the f-MWCNTs, which indicated that the number of defects increased in the nanocomposites.

Raman spectroscopy can be used for the determination of phase transitions in carbon nanotubes (Liu et al., 2012a, 2012b) and to investigate the crystallinity of nanomaterials

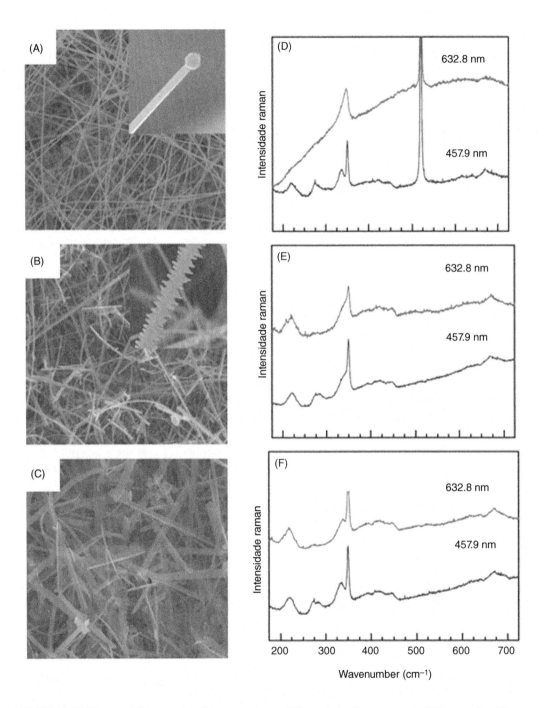

FIGURE 3.16 SEM images of (A) nanowires, (B) nanocombs, and (C) nanobelts. Raman spectra of (D) nanowires, (E) nanocombs, and (F) nanobelts excited with 1.96 and 2.71 eV. *SEM,* Scanning electron microscopy. *Adapted from Kim JH, et al: Raman spectroscopy of ZnS nanostructures,* J Raman Spectrosc *43(7):906–910, 2012.*

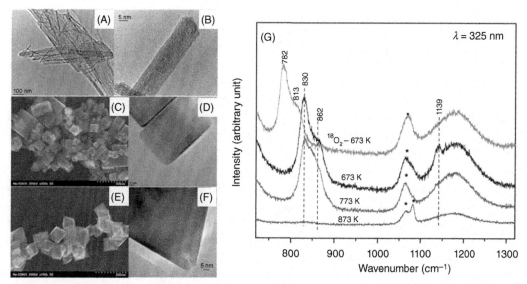

FIGURE 3.17 TEM images of nanorods (A and B), nanocubes (C and D), and nanooctahedra (E and F). (G) Raman spectra of nanorods reduced with thermal treatment and followed by O_2 adsorption. The asterisks indicate the presence of nitrate/phosphate. *TEM,* transmission electron microscopy. *Adapted from Wu ZL, et al: Probing defect sites on CeO2 nanocrystals with well-defined surface planes by Raman spectroscopy and O-2 adsorption,* Langmuir *26(21):16595–16606, 2010.*

and nanostructured materials. In the case of nanorods of zinc oxide (ZnO) and hematite (α-Fe$_2$O$_3$), Najjar et al. (2011) observed a large and visible band between 300 and 600 cm^{-1} when compared to ZnO without nanoparticles. This suggests a disruption of the symmetry of the crystal due to defects or impurities. By coupling atomic force microscopy (AFM) to Raman spectroscopy, it was possible to determine the orientation, crystallinity, and composition of a single nanorod of α-Fe$_2$O$_3$. Raman spectroscopy was also used to characterize cubic and hexagonal crystalline phases of nanorods of gadolinium oxide doped with europium (Gd$_2$O$_3$:Eu^{3+}) (Dhananjaya et al., 2011). While the cubic phase exhibits several bands located at approximately 137, 308, 389, and 489 cm^{-1}, the hexagonal phase shows a band only at 359 cm^{-1}.

Regarding nanostructured materials, Zanfolim et al. (2010) applied Raman spectroscopy to characterize ZnPc films grown using the PVD technique. The Raman spectra of the powder ZnPc show two bands (420 and 717 cm^{-1}) that are not present in the 40- and 400-nm films, and the other bands have different intensities. Furthermore, the similarity between the spectra of both films indicates that they have the same crystal structure, which is different from that present in the powder. Raman spectroscopy has also been applied to study the effect of temperature on the crystallinity of films. Raman spectra of the 40- and 400-nm-thick films were obtained before and after heating. The results indicated that the crystallinity of the 400-nm films is affected by heating but that the crystallinity of the 40-nm films is not affected. The Raman spectroscopy results also showed that the gasoline vapor chemically interacts with phthalocyanine, and it is possible to apply this film as a transducer in gas sensors.

Other nanomaterial properties can be studied using Raman spectroscopy, such as the mechanical properties of silicon nanowires (SiNWs) (Khachadorian et al., 2011) and the size distribution of nanocrystals. Georgescu et al. (2012) worked with titanium oxide aerogel (TiO_2). Their calculations were performed using the peak position and the full width at half-maximum for three major bands at 144, 398, and 638 cm^{-1}. The band at 144 cm^{-1} was more susceptible to the changes in the crystalline structure of the sample, and the band at 398 cm^{-1} was more sensitive to the variations in the size of the nanocrystals, shifting to shorter wavelengths as the size of the nanocrystals increased. Regarding the full width at half-maximum, the most sensitive band is the one at 638 cm^{-1}, in which the width tends to decrease with an increase in the nanocrystal size.

Raman spectroscopy can be used to assess the doping of conducting polymers. Zucolotto et al. (2006) investigated thin films of polyaniline (PANI) with three phthalocyanines, iron (FeTsPc), nickel (NiTsPc), and copper (CuTsPc), grown using the LbL technique. The Raman spectra showed that the primary doping was due to protonation and the secondary doping was due to structural and conformational changes in the PANI chain. The interactions between the phthalocyanine SO_3^- groups and PANI NH groups allow the growth of the LbL films. The similarity between the spectra indicated that the phthalocyanine did not induce the secondary doping. The work of Silva et al. (2012) using Raman spectroscopy on hybrid thin films of a semiconductor oxide (hexaniobate nanospirals) and PANI indicated that the semiconductor oxide is capable of inducing secondary doping in PANI to further improve its conductive property.

The photodegradation of Congo red (CR) can also be studied using Raman spectroscopy. CR was adsorbed onto the last layer of thin films with titanium dioxide nanoparticles (TiO_2) and polyelectrolytes, such as PAH and sodium polystyrene sulfonate (SPS), with $PAH/PSS/TiO_2(PSS/TiO_2)_5$ architecture (Sansiviero et al., 2011). The film was irradiated with UV light for 24 h, and Raman spectra were obtained, which revealed the emergence of new bands attributed to the CR oxidation process. The CR characteristic bands at 1595 cm^{-1} (phenyl ring), 1457 cm^{-1} (C=C stretching), and 1155 cm^{-1} appeared with very low intensity after the irradiation, which indicated photodegradation and photoisomerization. In addition, Raman spectra of commercially purchased TiO_2 in the anatase phase and of the LbL film $PAH/PSS/TiO_2(PSS/TiO_2)_5$ were obtained. The results indicated that the bands present in the film are broader and shifted to the red relative to pure TiO_2. This phenomenon is attributed to a disruption in the dynamic selection rule of phonons, which is attributed to ordered systems.

4 Surface-Enhanced Raman Scattering

Nanostructured materials show comparative advantages in efficiency and stability when compared to certain properties related to the volume (bulk) of the material. In the previous sections, UV–vis and infrared (IR) absorption spectroscopy technique applications, in addition to Raman scattering, to obtain molecular information about nanostructured materials with an emphasis on electronic transitions, interactions, molecular orientation, and structural identification of specific analytes were shown. Raman scattering is not particularly destructive, but it has a relatively small cross-section compared to the absorption and emission processes, which complicates the characterization of dilute solutions or nanosized structures such as ultrathin films. However, these systems can be mixed with metallic nanostructures, that is, solutions diluted in metal colloids and thin films deposited on roughened metal surfaces, to enhance the Raman signal by a factor of up to 10^7. This spectroscopy technique is known as SERS. The grounds and selection rules of SERS spectroscopy, in addition to the preparation of metallic nanoparticles, have been widely described (Aroca, 2006; Ru and Etchegoin, 2009). Jensen et al. (2008) published a review in which they mention four enhancement mechanisms (CHEM, *resonance*, CT, IN) based on resonant and nonresonant processes.

The development of a wide range of nanostructured materials requires control strategies for the chemical surface and nanoparticle synthesis in the search for specific properties (adsorption, electron transfer, stability, etc.). Thus, a wide range of nanostructured materials can be characterized using SERS spectroscopy when (1) they are mixed with metal nanoparticles; (2) they are deposited on roughened metal surfaces; (3) they are covered with evaporated metal films; and (4) they are dipped in metal ion solutions to induce chemical reduction with laser excitation. In this section, SERS characterization is addressed with a focus on the molecular identification of analytes on metal nanostructures or incorporated in nanostructured materials in sensor units, electronic circuits, biological materials, and historical heritage.

4.1 Sensor Units and Metallic Nanostructures

Several nanostructured systems are manufactured as multicomponent systems and mix organic and inorganic materials, such as natural rubber/AuNPs (Cabrera et al., 2012), SWNTs/Ag nanoparticles (AgNPs) or Au/polyethylene glycol (PEG) (Wang et al., 2012), and AuNPs/graphene oxide (Au–GO) (Huang et al., 2010). The immobilization of specific compounds improves the analyte adsorption properties on the surface. In this context, the SERS technique allows for the identification of the nanomaterials of the sensory unit and/or analytes detected at low concentrations, which also allows for the inference of the molecular orientation in the metal structure.

To detect the drug methylene blue (MB) and highlight its interactions with phospholipid structures, such as simple biological membrane models, Aoki et al. (2010) developed a sensory unit by depositing layers of phospholipids on interdigitated electrodes of Pt using the LbL technique. FTIR measurements of the LbL layers showed shifts in the signals (compared to the cast deposit film) for cardiolipin (CLP) phosphate groups (v_{ass} PO_2: $1235 \rightarrow 1215$ cm^{-1}) and the amino groups of PAH (δ NH_3: $1514 \rightarrow 1536$ and $1613 \rightarrow 1638$ cm^{-1}). Impedance measurements allow for detection of MB at low concentrations (10^{-7}, 10^{-9}, and 10^{-11} M). These results allowed us to infer that the adsorption of MB on the CLP phospholipid surface is favored by the electrostatic interaction of the phosphate groups. To obtain structural information using the SERS technique, the CLP

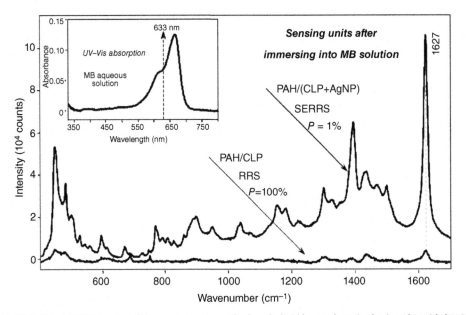

FIGURE 3.18 RRS and SERRS spectra of the sensory units with phospholipid layers deposited using LbL with (HAP/CLP)₅ and [PAH/(CLP AgNP)]₅ architectures, respectively. The spectra were collected with different laser intensities (*P*) at 633 nm. RRS refers to a region containing MB domains, while SERRS refers to a region without these domains (micrometer size). The inset shows the UV–vis absorption spectrum of the aqueous solution of MB (5 mM). *AgNP*, Ag nanoparticle; *CLP*, cardiolipin; *LbL*, layer-by-layer; *MB*, methylene blue; *PAH*, poly(allylamine hydrochloride); *RRS*, resonant Raman spectra; *SERRS*, resonant Raman spectra and surface-enhanced Raman scattering; *UV–vis*, ultraviolet–visible. *Adapted from Aoki PHB, et al: Coupling surface-enhanced resonance Raman scattering and electronic tongue as characterization tools to investigate biological membrane mimetic systems,* Anal Chem *82:3537–3546, 2010.*

was mixed with AgNPs, which alter the electrical response in the impedance measurements of the sensor based on the change in the charge balance on the surface. Fig. 3.18 shows the resonant Raman spectra (RRS) and SERRS (RRS + SERS) of MB adsorbed on a sensory unit with and without AgNPs, respectively. MB shows light absorption in the visible range, and a process in resonance with the 633-nm laser excitation. The most intense SERRS bands appear at 1623 cm^{-1} (ν C=C), 1500 and 1302 cm^{-1} (δ_{ring} C–C), 1393 cm^{-1} (ν C–N=C), and 1156 cm^{-1} (δ CH). The level of MB aggregation can be determined from the relative intensity of the bands between 450 and 500 cm^{-1} (monomers) and 480 cm^{-1} (aggregate structures). The monomeric species adsorbed in the sensory unit are detected only using SERRS (RRS is not able to detect), and it is possible to obtain an enhancement factor of the Raman signal at a value of 2×10^3 (1627-cm^{-1} band).

Metallic nanostructures were used for SERS information on electrodes (Dick et al., 2002), self-assembled monolayers (Orendorff et al., 2005), and glucose detection (Yonzon et al., 2004). One example is the metal film over nanosphere (MFON)-type nanostructures, which are nanospheres coated with evaporated metal. In the work by Zhang et al. (2006), AgFON coated with a layer of silica was used for the detection of anthrax [calcium dipicolinate (CaDPA)] in bacteria spores (*Bacillus subtilis*). Glass nanospheres

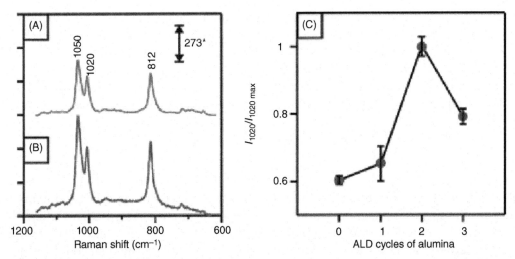

FIGURE 3.19 CaDPA SERS spectra at a concentration of 2 × 10⁻⁵ M in (A) AgFON–silica substrate (two alumina deposit cycles) and (B) AgFON nanoparticles. (C) SERS intensity at 1020 cm⁻¹ as a function of the thickness of alumina on AgFON. *ALD*, Atomic layer deposition; *CaDPA*, calcium dipicolinate; *SERS*, surface-enhanced Raman scattering. *Adapted from Zhang X, et al: Ultrastable substrates for surface-enhanced Raman spectroscopy: Al2O3 overlayers fabricated by atomic layer deposition yield improved anthrax biomarker detection, J Am Chem Soc 128:10304–10309, 2006.*

(18 mm diameter) were deposited on a surface and coated with two films, a vacuum-evaporated Ag film (pressure of 10^{-6} Torr) and an alumina film (1 nm) deposited by atomic layers in the gas phase [atomic layer deposition (ALD)]. Fig. 3.19 shows the CaDPA SERS spectra with AgFON and AgFON–silica nanostructures in addition to the SERS intensity as a function of the silica coating. The results indicated that the most relevant CaDPA SERS bands are at 1020 cm⁻¹ (v ring) and 812 cm⁻¹. In this case, the SERS technique was used for the analytical identification of CaDPA and for evaluating the adsorption on the AgFON–silica nanostructure. The intensity of the 1020-cm⁻¹ band was used to determine the presence of CaDPA at concentrations from 10^{-14} to 10^{-12} M, and an adsorption constant of $K_{spore} = 9.0 \times 10^{13}$ M⁻¹ was obtained.

Another possibility for SERS enhancement is the use of transition metals such as Fe, Co, Ni, Ru, Rh, Pd, and Pt. Tian et al. (2002) estimated the pyridine amplification factors on the roughened surface of seven transition metals (Fe, Co, Ni, Ru, Rh, Pd, Pt) with variations from 10^1 to 10^4 depending on the surface metal and pretreatment. The SERS amplification factors were calculated using the ratio between the SERS intensity and the number of pyridine molecules, both those adsorbed on the surface and in the solution: FA = $(I_{surf}/C_{surf})/(R_{solution}/N_{solution})$; the intensities correspond to the vibrational mode, v_1 (ring breathing, 1010 cm⁻¹). Abdelsalam et al. (2007) worked with electrodeposited substrates of Pt and Pd (0.25 and 0.05 V) on polystyrene spheres (self-assembly) and compared them to the evaporated surfaces of the metals. It was possible to obtain SERS enhancement with the 633-nm laser for benzenethiol (characteristic bands at 2600, 1571, 1071, 1021, and 996 cm¹), and the absence of the S–H stretching band at 2600 cm⁻¹ indicated a chemical bond with the metal (Pd–S). The calculations of the SERS enhancement (at 1571 cm^{-1}) of benzenethiol in the presence of Pt and Pd were obtained using the model created by Tian et al. (2002) for a 600-nm sphere assuming a coating of 0.45 nmol/cm^2 benzenethiol on the metal surface (roughness of 3.0), and the obtained enhancement values were 1800 (Pt) and 550 (Pd). These results were compared to benzenethiol adsorbed on Au (Abdelsalam et al., 2005) with a spectral correlation profile, and the amplification is greater when Au is used as the metal.

4.2 Electronic Circuits

The deformed silicon has a different crystal orientation and reduced electron transfer resistance compared to silicon, and it is used with metal oxide semiconductors to improve the speed in integrated circuits. The region in the integrated circuit where the device switches the flow of current is nanosized (~50 nm) (Leong et al., 2004). A study by Hayazawa et al. (2005) reported a SERS spectrum for a deformed silicon layer on a semiconductor substrate based on a coating of an evaporated Ag film. Fig. 3.20 shows a

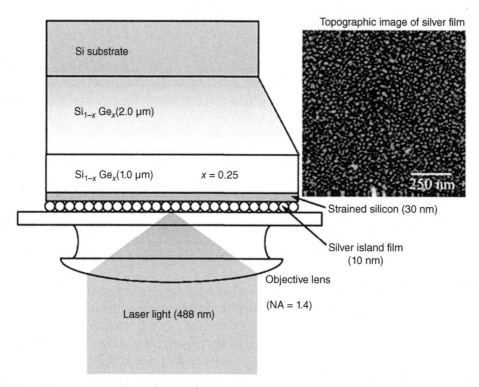

FIGURE 3.20 SERS microscopy diagram for identifying the deformed silicon layer and the atomic force microscopy (*AFM*) topographic image of the evaporated Ag film. *SERS*, Surface-enhanced Raman scattering. *Adapted from Hayazawa N, Motohashi M, Saito Y, Kawata S: Highly sensitive strain detection in strained silicon by surface-enhanced Raman spectroscopy, Appl Phys Lett 86:263114 (3 pp), 2005.*

diagram of the experimental system in which the deformed silicon layer (30-nm thick) was deposited on a Ge-doped substrate, and Ag was evaporated onto it (10-nm thick and 10^{-6} Torr vacuum). SERS signals of the material were obtained at 504.9 and 515.3 cm^{-1}, and they were attributed to the Si–Si vibrational modes of silicon and deformed silicon, respectively. In this case, the SERS detection sensitivity allowed for an accurate analysis to discriminate between the two species of silicon.

With SERS microscopy, it is difficult to have semiconductor selectivity in the region where the electronic device switches the current flow (approximately nanometer). To perform a selective analysis of the surface topography, AFM can be used with metal tips or metal coating to achieve a highly localized enhancement of the SERS signal for deformed silicon. This technique is called tip-enhanced Raman scattering (TERS). Tarun et al. (2009) published a review assessing the TERS technique for the characterization of semiconductors used in integrated circuits.

4.3 Biological Materials

Biological materials are composed of various structures, making them highly heterogeneous systems. For instance, biofilms are a highly heterogeneous matrix formed by polysaccharides, proteins, and other biopolymers (Branda et al., 2005). In biology, biofilms are defined as bacteria fixed on a surface due to adhesion properties. Schmid et al. (2008) worked with alginate fibers in water to form a hydrogel that protects the biofilm by closing the pores to the passage of water in bacteria. Alginate fibers were deposited on glass and soaked overnight in a 2% Ag nitrate solution. The SERS measurements were obtained using the laser ablation technique in which the laser used for Raman measurements causes the chemical reduction of $Ag^{1+} \rightarrow Ag^0$ directly in the sample. The Raman bands at 1413 cm^{-1} (Na alginate) and 1433 cm^{-1} (Ca alginate) were attributed to the stretching modes of COO$^-$, and the frequency shifting may be associated with the monodentate (Na) and bidentate (Ca) union with the carbonyl group of the alginate. Fig. 3.21 shows the Raman and SERS spectra of alginate fibers deposited on the surface. In this case, the SERS amplification effect is not necessary to identify the alginate, but it could be used to obtain information about the conformational differences in the polysaccharides (Na and Ca alginate), that is, the orientation of the C–C backbone (~810 cm^{-1}) and carbonyl groups (ν C–O at 810 cm^{-1}, δ C–C–O at 888 cm^{-1}) in the metal structure. The TERS technique was used to improve the selectivity and to obtain information from the surface about the alginate fibers. TERS studies were also conducted on amyloid fibers (Hermann et al., 2011) and cytochrome *c* (Yeo et al., 2008).

The preparation of reproducible and organized nanostructures is essential for obtaining materials with light scattering properties. Some naturally obtained materials have impressive optical effects. Butterflies use several layers of cuticles and air to produce the blue color on their wings, and some insects use matrices of elements to reduce the reflectivity of the compounds in their eyes (Vukusic and Sambles, 2003). Tan et al. (2012) used the wings of Lepidoptera (butterflies and moths) as highly organized substrates and made

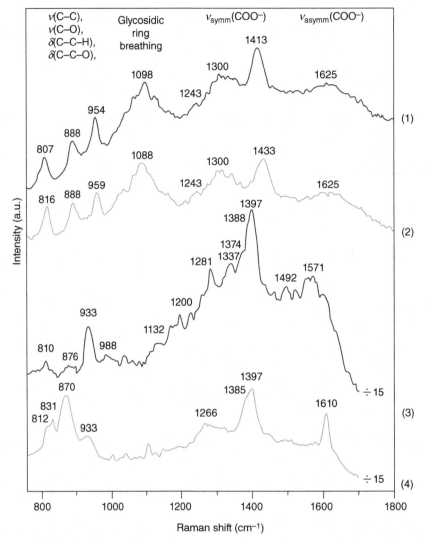

FIGURE 3.21 Alginate fibers of Na and Ca. Raman spectra of Na alginate (1) and Ca alginate (2). SERS spectra of Na alginate (3) and Ca alginate (4). *SERS*, Surface-enhanced Raman scattering. *Adapted from Schmid T, et al: Tip-enhanced Raman spectroscopy and related techniques in studies of biological materials,* Anal Bioanal Chem *391(1907–1916), 2008.*

an electrolytic coating with Cu (10 min) to obtain a highly organized structure suitable for amplification of the Raman signal (SERS). The substrates coated with Cu were tested with rhodamine 6G, and a reproducible SERS spectrum was obtained at a concentration of 10^{-5} M. The intensity of the band at 1650 cm^{-1} was used to compare the different nanostructures.

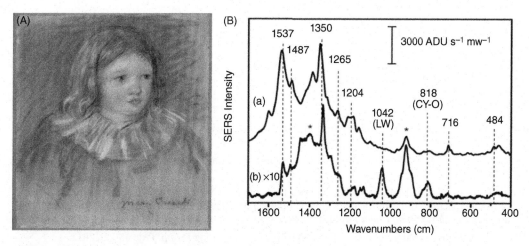

FIGURE 3.22 (A) Pastel by the artist Mary Cassatt ("Sketch of Margaret Sloane, Looking Right"). (B) SERS spectra of samples (a) pastel number 7 and (b) face in the painting. The lead white (*LW*) dye and chrome yellow orange dye (*CY-O*) signals are marked on the spectrum. *SERS*, Surface-enhanced Raman scattering. *Adapted from Brosseau CL, et al: Surface-enhanced Raman spectroscopy: a direct method to identify colorants in various artist media, Anal Chem 81(7443–7447), 2009.*

4.4 Paintings and Textiles of Historical Value

An alternative to characterizing historical paintings and textiles is to identify the dyes present in the sample using metal nanoparticles to generate an amplification of the Raman scattering (SERS) and suppression of the pigment fluorescence. SERS identification of flavonoids in textiles (Jurasekova et al., 2008) and of various dyes, such as alizarin, purpurin, carminic acid, hematoxylin, fisetin, quercitrin, quercetin, rutin, and morin, has been conducted (Leona et al., 2006). Brosseau et al. (2009) conducted a comparative SERS study of a painting by artist Mary Cassatt (pastel studio: "Sketch of Margaret Sloane, Looking Right") and the pastels ("pencils") she used (Boston Museum of Fine Arts). Fig. 3.22 shows the painting by the artist Mary Cassatt and the SERS spectra for the pastel sample number 7 and the red pigment obtained from the girl's face. The SERS signals of the lead white (LW) dye and chrome yellow-orange (CY-O) are marked in the spectra, but the red pigment was not identified and may be related to a monoazo dye type. In general, various dyes have been identified in pastels. SERS identification of the dyes in the work and in the pastels is important to evaluate the authenticity of the work.

List of Symbols

α Absorption coefficient
μ Molecular dipole moment
A   Absorbance
c   Concentration
E   Electric field

$E_{//}$ Parallel electric field
$E_{\perp}$ Perpendicular electric field
I Intensity
l Light path

References

Abdelsalam ME, et al: Electrochemical SERS at a structured gold surface, *Electrochem Commun* 7:740–744, 2005.

Abdelsalam ME, et al: SERS at structured palladium and platinum surfaces, *J Am Chem Soc* 129:7399–7406, 2007.

Alessio P, Luz Rodriguez-Mendez M, De Saja Saez JA, Leopoldo Constantino CJ: Iron phthalocyanine in non-aqueous medium forming layer-by-layer films: growth mechanism, molecular architecture and applications, *Phys Chem Chem Phys* 12(16):3972–3983, 2010.

Aljada M, et al: Structured-gate organic field-effect transistors, *J Phys D Appl Phys* 45(22), 2012.

Aoki PHB, et al: Layer-by-layer technique as a new approach to produce nanostructured films containing phospholipids as transducers in sensing applications, *Langmuir* 25(4):2331–2338, 2009a.

Aoki PHB, et al: Taking advantage of electrostatic interactions to grow Langmuir–Blodgett films containing multilayers of the phospholipid dipalmitoylphosphatidylglycerol, *Langmuir* 25(22):13062–13070, 2009b.

Aoki PHB, et al: Coupling surface-enhanced resonance Raman scattering and electronic tongue as characterization tools to investigate biological membrane mimetic systems, *Anal Chem* 82:3537–3546, 2010.

Aoki PHB, et al: Spray layer-by-layer films based on phospholipid vesicles aiming sensing application via e-tongue system, *Mater Sci Eng C Mater Biol Appl* 32(4):862–871, 2012.

Aroca R: *Surface-enhanced vibrational spectroscopy*, Chichester, 2006, John Wiley & Sons Ltd.

Blodgett KB: Films built by depositing successive monomolecular layers on a solid surface, *J Am Chem Soc* 57:1007–1022, 1935.

Bokobza L, Zhang J: Raman spectroscopic characterization of multiwall carbon nanotubes and of composites, *Express Polym Lett* 6(7):601–608, 2012.

Born M, Wolf E: *Principles of optics*, Oxford, 1987, Pergamon Press.

Bradshaw AM, Schweizer E: *Spectroscopy of surfaces*, Toronto, 1988, John Wiley & Sons.

Branda SS, Vik A, Friedman L, Kolter R: Biofilms: the matrix revisited, *Trends Microbiol* 13:20–26, 2005.

Brosseau CL, et al: Surface-enhanced Raman spectroscopy: a direct method to identify colorants in various artist media, *Anal Chem* 81:7443–7447, 2009.

Cabrera FC, et al: Portable smart films for ultrasensitive detection and chemical analysis using SERS and SERRS, *J Raman Spectrosc* 43:474–477, 2012.

Centurion LMPC, Moreira WC, Zucolotto V: Tailoring molecular architectures with cobalt tetrasulfonated phthalocyanine: immobilization in layer-by-layer films and sensing applications, *J Nanosci Nanotechnol* 12(3):2399–2405, 2012.

Chai F, et al: Colorimetric detection of Pb2+ using glutathione functionalized gold nanoparticles, *ACS Appl Mater Interfaces* 2(5):1466–1470, 2010.

Chang HC, et al: Preparation and electrochemical characterization of NiO nanostructure–carbon nanowall composites grown on carbon cloth, *Appl Surf Sci* 258(22):8599–8602, 2012.

Cheng X, et al: Characterization of multiwalled carbon nanotubes dispersing in water and association with biological effects, *J Nanomater* 2011, 2011.

Correia FC, Wang SH, Péres LO, Caseli L: Langmuir and Langmuir–Blodgett films of a quinoline-fluorene based copolymer, *Colloids Surf A Physicochem Eng Aspects* 394(0):67–73, 2012.

Cotrone S, et al: Phospholipid film in electrolyte-gated organic field-effect transistors, *Org Electron* 13(4):638–644, 2012.

Crespilho FN, et al: The origin of the molecular interaction between amino acids and gold nanoparticles: a theoretical and experimental investigation, *Chem Phys Lett* 469(1–3):186–190, 2009.

de Oliveira RF, De Moraes ML, Oliveira ON, Ferreira M: Exploiting cascade reactions in bienzyme layer-by-layer films, *J Phys Chem C* 115(39):19136–19140, 2011.

Debe MK: Optical probes of organic thin films photons-in and photons-out, *Prog Surf Sci* 24(1–4):1–282, 1987.

Decher G: Fuzzy nanoassemblies: toward layered polymeric multicomposites, *Science* 277(5330):1232–1237, 1997.

Del Cano T, Duff J, Aroca R: Molecular spectra and molecular organization in thin solid films of bis(neopentylimido) perylene, *Appl Spectrosc* 56(6):744–750, 2002.

Del Cano T, et al: Molecular stacking and emission properties in Langmuir–Blodgett films of two alkyl substituted perylene tetracarboxylic diimides, *Org Electron* 5(1–3):107–114, 2004.

Dhananjaya N, et al: Hydrothermal synthesis, characterization and Raman studies of Eu3+ activated Gd2O3 nanorods, *Physica B Condens Matter* 406(9):1639–1644, 2011.

Dick LA, Mcfarland AD, Haynes CL, Duyne RPV: Metal film over nanosphere (MFON) electrodes for surface-enhanced Raman spectroscopy (SERS): improvements in surface nanostructure stability and suppression of irreversible loss, *J Phys Chem B* 106:853–860, 2002.

Ferreira M, et al: Langmuir and Langmuir–Blodgett films of poly 2-methoxy-5-(*n*-hexyloxy)-*p*-phenylenevinylene, *Langmuir* 19(21):8835–8842, 2003.

Georgescu D, et al: Experimental assessment of the phonon confinement in TiO2 anatase nanocrystallites by Raman spectroscopy, *J Raman Spectrosc* 43(7):876–883, 2012.

Graupner R: Raman spectroscopy of covalently functionalized single-wall carbon nanotubes, *J Raman Spectrosc* 38(6):673–683, 2007.

Hayazawa N, Motohashi M, Saito Y, Kawata S: Highly sensitive strain detection in strained silicon by surface-enhanced Raman spectroscopy, *Appl Phys Lett* 86:263114 (3 pp), 2005.

He W, et al: Design of AgM bimetallic alloy nanostructures (M = Au, Pd, Pt) with tunable morphology and peroxidase-like activity, *Chem Mater* 22(9):2988–2994, 2010.

Heriot SY, Zhang H-L, Evans SD, Richardson TH: Multilayers of 4-methylbenzenethiol functionalized gold nanoparticles fabricated by Langmuir–Blodgett and Langmuir–Schaefer deposition, *Colloids Surf A Physicochem Eng Aspects* 278(1–3):98–105, 2006.

Hermann P, Fabian H, Naumann D, Hermelink A: Comparative study of far-field and near-field Raman spectra from silicon-based samples and biological nanostructures, *J Phys Chem C* 115:24512–24520, 2011.

Hu X, Ji J: Covalent layer-by-layer assembly of hyperbranched polyether and polyethyleneimine: multilayer films providing possibilities for surface functionalization and local drug delivery, *Biomacromolecules* 12(12):4264–4271, 2011.

Huang J, et al: Nanocomposites of size-controlled gold nanoparticles and graphene oxide: formation and applications in SERS and catalysis, *Nanoscale* 2:2733–2738, 2010.

Jans H, Huo Q: Gold nanoparticle-enabled biological and chemical detection and analysis, *Chem Soc Rev* 41(7):2849–2866, 2012.

Jensen L, Aikens CM, Schatz GC: Electronic structure methods for studying surface-enhanced Raman scattering, *Chem Soc Rev* 37:1061–1073, 2008.

Jurasekova Z, Domingo C, Garcia-Ramos JV, Sanchez-Cortes S: In situ detection of flavonoids in weld-dyed wool and silk textiles by surface-enhanced Raman scattering, *J Raman Spectrosc* 39:1309–1312, 2008.

Kam AP, Aroca R, Duff J: Perylene tetracarboxylic-phthalocyanine mixed thin solid films. Surface-enhanced resonance Raman scattering imaging studies, *Chem Mater* 13(12):4463–4468, 2001.

Khachadorian S, et al: High pressure Raman scattering of silicon nanowires, *Nanotechnology* 22(19):195707, 2011.

Kim D, et al: Raman characterization of thermal conduction in transparent carbon nanotube films, *Langmuir* 27(23):14532–14538, 2011.

Kim JH, et al: Raman spectroscopy of ZnS nanostructures, *J Raman Spectrosc* 43(7):906–910, 2012.

Kobayashi Y, Takagi D, Ueno Y, Homma Y: Characterization of carbon nanotubes suspended between nanostructures using micro-Raman spectroscopy, *Physica E Low Dimensional Syst Nanostruct* 24(1–2):26–31, 2004.

Kuila T, et al: Chemical functionalization of graphene and its applications, *Prog Mater Sci* 57(7):1061–1105, 2012.

Langmuir I: The constitution and fundamental properties of solids and liquids. II. Liquids, *J Am Chem Soc* 39:1848–1906, 1917.

Leona M, Stenger J, Ferloni E: Application of surface-enhanced Raman scattering techniques to the ultrasensitive identification of natural dyes in works of art, *J Raman Spectrosc* 37:981–992, 2006.

Leong M, et al: Silicon device scaling to the sub-10-nm regime, *Science* 306:2057–2060, 2004.

Li J, et al: Rolling circle amplification combined with gold nanoparticle aggregates for highly sensitive identification of single-nucleotide polymorphisms, *Anal Chem* 82(7):2811–2816, 2010a.

Li L, Li B, Cheng D, Mao L: Visual detection of melamine in raw milk using gold nanoparticles as colorimetric probe, *Food Chem* 122(3):895–900, 2010b.

Linic S, Christopher P, Ingram DB: Plasmonic-metal nanostructures for efficient conversion of solar to chemical energy, *Nat Mater* 10(12):911–921, 2011.

Liu, B, Han, G, et al.: Shell thickness-dependent Raman enhancement for rapid identification and detection of pesticide residues at fruit peels, *Anal Chem* 84(1):255–261, 2011.

Liu B, et al: Shell thickness-dependent Raman enhancement for rapid identification and detection of pesticide residues at fruit peels, *Anal Chem* 84(1):255–261, 2012a.

Liu DD, et al: In situ Raman and photoluminescence study on pressure-induced phase transition in C-60 nanotubes, *J Raman Spectrosc* 43(6):737–740, 2012b.

Mantha S, et al: Renewable nanocomposite layer-by-layer assembled catalytic interfaces for biosensing applications, *Langmuir* 26(24):19114–19119, 2010.

Najjar S, et al: Characterization of single transition metal oxide nanorods by combining atomic force microscopy and polarized micro-Raman spectroscopy, *Chem Phys Lett* 514(1–3):128–133, 2011.

Orendorff CJ, Gole A, Sau TK, Murphy CJ: Surface-enhanced Raman spectroscopy of self-assembled monolayers: sandwich architecture and nanoparticle shape dependence, *Anal Chem* 77:3261–3266, 2005.

Pardo H, et al: Raman characterization of bulk ferromagnetic nanostructured graphite, *Physica B Condens Matter* 407(16):3206–3209, 2012.

Pasquini D, et al: Surface morphology and molecular organization of lignins in Langmuir–Blodgett films, *Langmuir* 18(17):6593–6596, 2002.

Pavinatto FJ, et al: Optimized architecture for tyrosinase-containing Langmuir–Blodgett films to detect pyrogallol, *J Mater Chem* 21(13):4995–5003, 2011.

Pereira AA, et al: Lignin from sugar cane bagasse: extraction, fabrication of nanostructured films, and application, *Langmuir* 23(12):6652–6659, 2007.

Petrozzi S, Goudie S: *Practical instrumental analysis*, Wiley, 2013.

Rao AM, et al: Effect of van der Waals interactions on the Raman modes in single walled carbon nanotubes, *Phys Rev Lett* 86(17):3895–3898, 2001.

Roro KT, et al: Solar absorption and thermal emission properties of multiwall carbon nanotube/nickel oxide nanocomposite thin films synthesized by sol–gel process, *Mater Sci Eng B Adv Funct Solid State Mater* 177(8):581–587, 2012.

Ru EL, Etchegoin P: *Principles of surface enhanced Raman spectroscopy and related plasmonic effects*, Oxford, 2009, Elsevier.

Sansiviero MTC, Dos Santos DS, Job AE, Aroca RF: Layer by layer TiO2 thin films and photodegradation of Congo red, *J Photochem Photobiol A* 220(1):20–24, 2011.

Scarselli M, Castrucci P, De Crescenzi M: Electronic and optoelectronic nano-devices based on carbon nanotubes, *J Phys Condens Matter* 24(31):313202, 2012.

Schmid T, et al: Tip-enhanced Raman spectroscopy and related techniques in studies of biological materials, *Anal Bioanal Chem* 391:1907–1916, 2008.

Silva CHB, et al: Spectroscopic, morphological and electrochromic characterization of layer-by-layer hybrid films of polyaniline and hexaniobate nanoscrolls, *J Mater Chem* 22(28):14052–14060, 2012.

Skoog DA, Holler FJ, Crouch SR: *Principles of instrumental analysis*, Belmont, CA, 2007, Thomson Brooks/Cole.

Suzuki S, Hibino H: Characterization of doped single-wall carbon nanotubes by Raman spectroscopy, *Carbon* 49(7):2264–2272, 2011.

Tan Y, et al: High-density hotspots engineered by naturally piled-up subwavelength structures in three-dimensional copper butterfly wing scales for surface-enhanced Raman scattering detection, *Adv Funct Mater* 22:1578–1585, 2012.

Tarun A, Hayazawa N, Kawata S: Tip-enhanced Raman spectroscopy for nanoscale strain characterization, *Anal Bioanal Chem* 394:1775–1785, 2009.

Tian ZQ, Ren B, Wu DY: Surface-enhanced Raman scattering: from noble to transition metals and from rough surfaces to ordered nanostructures, *J Phys Chem B* 106:9463–9483, 2002.

Umut E, et al: Magnetic, optical and relaxometric properties of organically coated gold–magnetite (Au–Fe3O4) hybrid nanoparticles for potential use in biomedical applications, *J Magnetism Magn Mater* 324(15):2373–2379, 2012.

Volpati D, et al: Exploiting distinct molecular architectures of ultrathin films made with iron phthalocyanine for sensing, *J Phys Chem B* 112(48):15275–15282, 2008a.

Volpati D, Job AE, Aroca RF, Constantino CJL: Molecular and morphological characterization of bis benzimidazo perylene films and surface-enhanced phenomena, *J Phys Chem B* 112(13):3894–3902, 2008b.

Vukusic P, Sambles JR: Photonic structures in biology, *Nature* 424:852–855, 2003.

Wang X, et al: Noble metal coated single-walled carbon nanotubes for applications in surface enhanced Raman scattering imaging and photothermal therapy, *J Am Chem Soc* 134:7414–7422, 2012.

Wu ZL, et al: Probing defect sites on CeO2 nanocrystals with well-defined surface planes by Raman spectroscopy and O-2 adsorption, *Langmuir* 26(21):16595–16606, 2010.

Xue C, Mirkin CA: pH-switchable silver nanoprism growth pathways, *Angew Chem Int Ed Engl* 46(12):2036–2038, 2007.

Yang Y, Jiang Y, Xu J, Yu J: Conducting polymeric nanoparticles synthesized in reverse micelles and their gas sensitivity based on quartz crystal microbalance, *Polymer* 48(15):4459–4465, 2007.

Yeo BS, et al: Tip-enhanced Raman spectroscopy can see more: the case of cytochrome c, *J Phys Chem C* 112:4867–4873, 2008.

Yonzon CR, et al: A glucose biosensor based on surface-enhanced Raman scattering: improved partition layer, temporal stability, reversibility, and resistance to serum protein interference, *Anal Chem* 76:78–85, 2004.

Zafer C, et al: New perylene derivative dyes for dye-sensitized solar cells, *Solar Energy Mater Solar Cells* 91(5):427–431, 2007.

Zanfolim AA, et al: Structural and electric-optical properties of zinc phthalocyanine evaporated thin films: temperature and thickness effects, *J Phys Chem C* 114(28):12290–12299, 2010.

Zhang X, et al: Ultrastable substrates for surface-enhanced Raman spectroscopy: Al2O3 overlayers fabricated by atomic layer deposition yield improved anthrax biomarker detection, *J Am Chem Soc* 128:10304–10309, 2006.

Zhang Y, et al: Shape effects of Cu2O polyhedral microcrystals on photocatalytic activity, *J Phys Chem C* 114(11):5073–5079, 2010.

Zhu Y, et al: Low temperature preparation of hollow carbon nano-polyhedrons with uniform size, high yield and graphitization, *Mater Chem Phys* 134(2–3):639–645, 2012.

Zucolotto V, et al: Nanoscale processing of polyaniline and phthalocyanines for sensing applications, *Sens Actuators B Chem* 113(2):809–815, 2006.

4

Dynamic Light Scattering Applied to Nanoparticle Characterization

Ana P. Ramos

UNIVERSITY OF SÃO PAULO, RIBEIRÃO PRETO, SÃO PAULO, BRAZIL

CHAPTER OUTLINE

1 Theory

1.1 Rayleigh Scattering

Different phenomena are observed when electromagnetic radiation hits a sample. These phenomena will depend on the radiation energy and the nature of the sample. For example, transitions between rotational levels are observed in the microwave region, whereas energy absorption in the ultraviolet region will result in transitions between electron levels. Therefore, electromagnetic radiation is one of the most important structural probes of matter. On photon incidence, energy can be transferred to (or absorbed by) different degrees of freedom of a molecule, that is, translational, rotational, vibrational, or electron. The resultant scattered radiation spectrum will show resonances at frequencies corresponding to these transitions.

When a particle is irradiated by a light source in the visible wavelength region, part of the light will be transmitted through the sample and part of the light can be absorbed by the sample. Furthermore, if the particle is small enough relative to the wavelength (λ) of the incident radiation (i.e., approximately $<\lambda/20$), the radiation will be scattered in different directions without changes in its energy or in the wavelength of the incident light. This phenomenon is known as Rayleigh scattering, or elastic scattering of light, and is related only to the translational and rotational degrees of freedom.

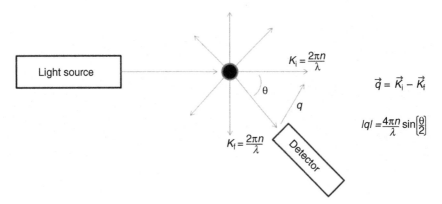

FIGURE 4.1 Schematic of light scattered by a particle whose size is equivalent to the wavelength of the incident radiation.

The electric field of electromagnetic radiation induces an oscillating polarization of electrons. Because of the appearance of an oscillating dipole, the particles become secondary sources of radiation with wavelengths the same as that of the incident radiation (Berne and Pecora, 2000). When this occurs, the intensity of the nonpolarized scattered light due to Rayleigh scattering by an individual particle is as follows:

$$I = I_0 \frac{1+\cos^2\theta}{2r^2}\left(\frac{2\pi}{\lambda}\right)^4\left(\frac{n^2-1}{n^2+2}\right)\left(\frac{d}{2}\right)^6 \tag{4.1}$$

where I_0 and λ are the intensity and wavelength of the incident radiation, respectively, r is the distance to the scattering center, θ is the angle at which the light is scattered, n is the refraction index, and d is the particle diameter. The unit term in $(1 + \cos^2\theta)$ represents the component of the scattered light that is vertically polarized; the $\cos^2\theta$ term represents the horizontally polarized component (Shaw, 1975). If the scattering centers are independent and randomly distributed, the probabilities of constructive and destructive interference between the scattered waves will be the same.

Fig. 4.1 is a schematic of the scattering of radiation with wavelengths of the same order of magnitude as the particle size in a typical light scattering experiment.

Note that vectors K_i and K_f have directions that are the same as those of the propagated incident light and of the radiation with the same wavelength that hits the detector, respectively. The scattering angle (θ) is defined by the detector position. The scattering vector (q) is geometrically defined as a function of θ, the refraction index of the medium (n), and the wavelength (λ) of the source of radiation.

If, for example, the radiation source is a laser, which is intrinsically monochromatic and coherent, the variations in the intensity of the scattered light will depend on time. The development of dynamic light scattering (DLS) grew with the development of laser manufacturing. The stronger the laser, the smaller are intensity variations of the incident light that can be measured. Because light is scattered in all directions, the scattering vector "q" is defined from the detector position. Typical values of "q" vary between

0.0042 and 0.031 nm^{-1} (for aqueous dispersions and λ = 514.5 nm) (Radeva, 2001). The intensity of Rayleigh scattering [Eq. (4.1)] also depends on the scattering angle. Therefore, we can compare only data obtained with the same angle. In commercial DLS equipment, the detector is usually positioned at 90 or 175 degrees from the laser source. This facilitates the comparison of data obtained from different instruments. Additionally, because of its dependence on sin θ, the scattering vector is at a maximum around those angles.

1.2 Brownian Movement and the DLS Technique

The DLS technique, also known as photon correlation spectroscopy, allows for the calculation of diffusion coefficients associated with the Brownian movement of particles dispersed in liquid media. This calculation is performed by irradiating a sample with a laser beam and analyzing the intensity fluctuations of the light scattered by the particles. Brownian movement is the movement of particles due to random collisions with molecules of liquid in their surroundings. An important feature of Brownian movement regarding the DLS technique is that small particles move faster than larger particles. In other words, the translational diffusion coefficient (D) of the particles is inversely related to their size, as shown by the Stokes–Einstein equation:

$$D = \frac{kT}{6\pi\eta R_{\mathrm{H}}} \tag{4.2}$$

where k is the Boltzmann constant, T is the temperature in Kelvin, η is the viscosity of the dispersing medium, and R_{H} is the hydrodynamic radius of the particle. This equation is valid for spherical particles that do not interact among themselves (Schmitz, 1990).

Because of Brownian movement, the distance between particles (i.e., scattering centers) varies with time, which creates constructive and destructive interferences in the intensity of the scattered light. Therefore, the fluctuation of scattered light intensity as a function of time reveals information on the velocity of the scattering centers, that is, the translational diffusion coefficient. Larger particles will cause smaller fluctuation rates in the scattered light, whereas smaller, faster particles will result in higher fluctuation rates, as shown in Fig. 4.2. These fluctuations in time of the scattered light intensity can then be converted to values for the diffusion coefficient and particle radius.

The correlator used in the DLS instrument will correlate the variations of intensity as a function of time and generate an autocorrelation function $G(t)$, defined by the following (Berne and Pecora, 2000):

$$G(t) \equiv \langle I(t_0)I(t_0 + t) \rangle \tag{4.3}$$

where $I(t_0)$ is the number of photons that reach the detector at time (t_0) and $I(t_0 + t)$ is the number of photons after a time interval (t). The rate of photocounting is proportional to the intensity of the light that reaches the detector. If the intensity decays as a simple exponential, the autocorrelation function given by a cumulative analysis (Schmitz, 1990) in a homodyne configuration (Berne and Pecora, 2000), which considers only scattered light

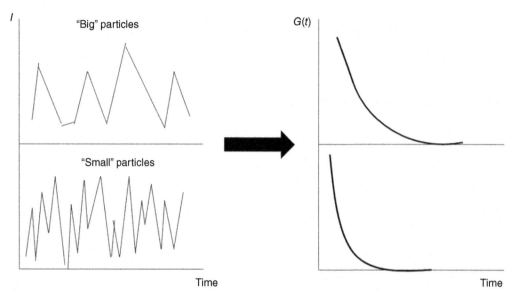

FIGURE 4.2 Fluctuations of intensity of scattered light as a function of time and their respective autocorrelation functions [$G(t)$].

that reaches the detector and discriminates the noise, is an exponential decay function with a characteristic relaxation time (τ):

$$G(t) = A + B e^{-2t/\tau} \tag{4.4}$$

where the factor 2 indicates the homodyne method, and A and B are parameters obtained from the fitting of the experimental points in the autocorrelation function. Here, the value of τ characterizes the autocorrelation function; as such, its value must vary during the data collection interval Δt (Schmitz, 1990). At long enough times, the system evolves to a state that is not correlated at all to the initial state.

Fig. 4.2 shows a schematic of the exponential profile of the curve generated for the autocorrelation function versus time after the intensity data obtained by the correlator are converted through the homodyne method (Berne and Pecora, 2000). Note that the auto-correlation curve starts at a maximum value [usually equal to (4.1)], where the correlation among the photons is at a maximum, and reaches zero, where the correlation is totally lost. The exponentials obtained from samples with small particles reach zero in less time than those with large particles, which indicates that the correlation is more rapidly lost because of the higher diffusion coefficients of the smaller particles.

The translational diffusion coefficient (D) can be calculated from the correlation function. For sufficiently diluted samples and more than one class of particle size, the autocorrelation function will be the sum of the decay times generated by each type of diffusion (Schmitz, 1990):

$$G(t) = A + [B_1 e^{-t/\tau_1} + B_2 e^{-t/\tau_2}]^2 \tag{4.5}$$

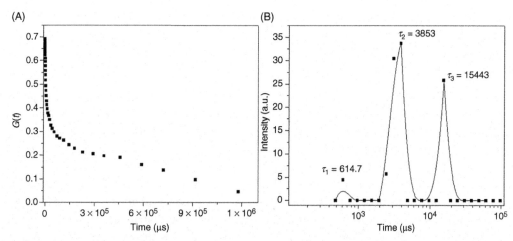

FIGURE 4.3 (A) Autocorrelation function [$G(t)$] and (B) spectrum of relaxation time (τ_i) versus time obtained for cellulose dispersion in a phosphate buffer (pH = 7.5).

Each decay time, or relaxation time (τ_i) in the spectrum, will correspond to a diffusion mode (Sedlák, 1999):

$$D=\left(\frac{1}{\tau_i}\right)q^{-2} \tag{4.6}$$

Fig. 4.3 shows an example of an autocorrelation function and the corresponding relaxation time spectrum (τ_i).

Fig. 4.3 shows that the autocorrelation function has an initial sharp decay because of the fast movement of the small particles, which causes a quick correlation loss. Fitting the function of Fig. 4.3 yields a multiexponential decay with three identified decay times (τ_1, τ_2, and τ_3). Each diffusion mode is associated with a relaxation time due to the different sizes of the cellulose fibers. The diffusion coefficient associated with each mode can be calculated from Eq. (4.6). The instruments use algorithms that fit the decay curves to monoexponential or multiexponential functions and estimate the distribution widths [polydispersity index (PDI)]. One of the most used algorithms to analyze polydispersed particles is Contin (Provder, 1987). Most instruments perform this analysis directly and apply the Stokes–Einstein equation [Eq. (4.2)] to calculate the hydrodynamic radius of a particle from the values of the diffusion coefficient. The analyses are easier to perform when the interactions and interparticle collisions are minimized. The number of collisions and charge effects are reduced by utilizing highly diluted samples and by adding an electrolyte to the medium, respectively.

Because the data registered by the instrument are values of the intensity of the scattered light as a function of time, the size distributions obtained by the DLS technique are shown as the relative intensity versus the diameter (or radius) of the particles. However, when this distribution has more than one peak, the relative contribution of each peak given by its intensity may not be representative of the number of particles that constitute the different size classes.

1.3 Analysis of the Distributions

The DLS technique shows the intensity of scattered light as a function of the particle size distribution. Particle size distribution can be converted into contributions per volume or number using the right algorithms. The conversion of relative distributions of intensity to volume can be performed using the Mie theory (Bohren and Huffman, 1983). For this, the refraction indices of the particles must be provided as inputs. This conversion will provide a more realistic distribution of the contribution of each peak. To better understand the concepts of distribution by intensity, volume, and number, consider a sample that has the same numbers of 5- and 50-nm diameter particles. The distribution by number will generate two peaks of the same relative percentage (i.e., a 1:1 ratio) centered at these sizes. If converted to volumes, the peak corresponding to the 50-nm particles will have a contribution 1000 times larger (i.e., a 1:1000 ratio) than that of the 5-nm particles because the volume of a sphere is proportional to (diameter)^3. The distribution by intensity obtained using the DLS technique shows that the contribution of the 50-nm particles' peak is at least 10^6 higher than that of the 5-nm particles' peak, considering that, according to the Rayleigh theory shown in Eq. (4.1), the intensity of the scattered light is proportional to (diameter)^6 (Malvern Instruments, 2003).

The average particle diameter can be determined when the system is monodisperse, that is, when the particle size distribution is narrow. For systems with more than one range of particle sizes, the use of available algorithms such as Contin may help in the size analysis. However, the analysis of systems with particles of different sizes using intensity distributions may not effectively correspond with the number of particles in each range.

The PDI of a sample, or its polydispersity, is an important parameter in DLS measurements and can be obtained from the widths of the particle size ranges. If the particle size distribution can be fitted to a Gaussian distribution, the PDI is associated with the standard deviation (σ) and the average hydrodynamic radius (R_H), as per the following (Koppel, 1972):

$$\mathrm{PDI} = \frac{\sigma^2}{R_H^2} \tag{4.7}$$

Thus, the PDI is simply the relative variance, whereas the polydispersity corresponds to the standard deviation (σ). The wider the distribution is, the more polydisperse the sample.

2 Applications

The value of the translational diffusion coefficient of a particle in a medium is dependent on how the particle moves in the medium. Therefore, the diameter obtained by DLS includes other molecules, even solvents, that surround the particle core and move together

(A)   (B)

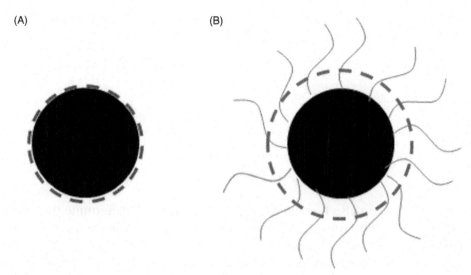

FIGURE 4.4 Particles with the same core size but with different surfaces are circumscribed by spheres of different diameters. Schematic representation of (A) a non-modified particle's surface and (B) a polymer-modified surface.

with the particle, that is, the so-called hydrodynamic diameter or radius. For example, the addition of electrolytes may change the hydrodynamic radius because of modifications in the ionic atmosphere that surrounds the particle through the reduction of the Debye length. This allows for the differentiation of the same-core-size particles but with surface differences, as shown in Fig. 4.4. The study of the surface modification of nanoparticles through hydrodynamic radius measurements (Gao et al., 2015; Martin et al., 2015) is one of the main applications of DLS.

Fig. 4.4A and B is schematic representation of the diameters of different spheres that circumscribe particles with the same core but different surfaces. The calculation by DLS of the size of the particle of Fig. 4.4A will be smaller than that of the particle of Fig. 4.4B. The conformations of the polymers linked to the particle surface, such as in Fig. 4.4B, will change the apparent size measured by DLS.

Polymer coating layers on metallic cores have seldom been observed by microscopy techniques, such as transmission electron microscopy (TEM), due to the low contrast of the polymer. However, these diameter variations can be measured by DLS. The silica layer coatings of Au and Ag metallic particles cannot be observed by high-resolution TEM (Jana et al., 2007). Therefore, the sizes measured by high-resolution TEM correspond only to the metallic cores, whereas the results obtained by DLS correspond to the total size of the metallic core and the coating layer. Table 4.1 shows a comparison of the Au and Ag particle sizes obtained by TEM and DLS. After considering for experimental deviations, the sizes of the metallic particles before surface modification obtained by both techniques were similar. However, after surface modifications, only DLS was capable of detecting the thickness of the coating layer covering the particles.

Table 4.1 Comparison of the Diameters of Au and Ag Nanoparticles Obtained by TEM and DLS, Before and After Coating With a Silica Layer

	Size, TEM (nm)	Size, DLS (nm)
Au before coating	6 ± 1	8 ± 2
Au after coating	7 ± 1	15 ± 2
Ag before coating	5 ± 2	7 ± 2
Ag after coating	6 ± 2	15 ± 4

DLS, Dynamic light scattering; *TEM*, transmission electron microscopy.
Adapted from Jana NR, Earhart C, Ying JY: Synthesis of water-soluble and functionalized nanoparticles by silica coating, *Chem Mater* 19(21):5074–5082, 2007.

Particle stability can also be studied by DLS. The ability to determine the minimum concentration of a stabilizing agent necessary to prevent particle aggregation is one of the applications of the technique (Guidelli et al., 2012; Tarasov et al., 2013). A periodic analysis may show aggregation, coagulation, or even sintering after thermal treatment, which would result in an increased average diameter due to several different factors, such as preparation time and reagent concentration (Guidelli et al., 2011). The formation of nanoscale complexes of different types of molecules and particles in solution can also be studied, as shown by Borissevitch et al. (2013). In this study, the formation of complexes between porphyrins and quantum points was characterized by DLS to explain observed fluorescence suppression phenomena. Glass and sol–gel transition studies in colloidal dispersions can also be performed (Van Mergen and Underwood, 1993) by considering the growth of the average diameters of colloids. DLS has also been used to study the incorporation efficiency of drugs by micelles (Yokoyama et al., 1998) and liposomes (Zaru et al., 2007) to produce controlled delivery systems. These studies observed increases in the average diameters of carriers after drug exposure.

Size analyses of nonspherical particles by DLS must be carefully performed. For spherical objects, the calculated diameter depends on the translational diffusion coefficient, which is then converted to hydrodynamic radius by the Stokes–Einstein equation (4.2). The diameter will always be approximated to that of a sphere diffusing in the same medium at the same velocity regardless of the particle shape. Changes in the particle shape that modify their translational diffusion coefficient will result in variations in the calculated diameter. For example, for a cylindrical object of 5000-nm length (L) and 200-nm diameter (d), the corresponding sphere with an equivalent translational diffusion coefficient has a diameter of approximately 670 nm, which does not accurately describe the actual dimensions of the object. Changes to L and d will alter the diffusion coefficient. While these particles could still be studied by DLS, the calculated hydrodynamic radii cannot be relied on to accurately represent particle geometries. The dimensional analysis of anisotropic particles can be performed using DLS with a depolarized laser source (Badaire et al., 2004). This isolates components of the incident light source into parallel and orthogonal directions to the particle main axis and thereby enables the calculation of the rotational diffusion coefficient related to the L and d dimensions. However, this analysis cannot be

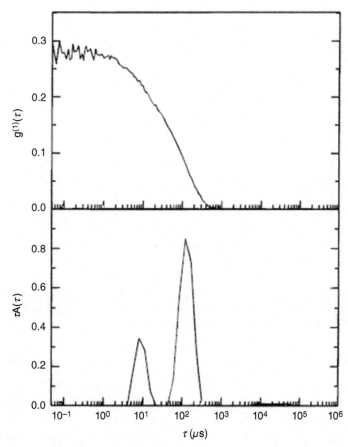

FIGURE 4.5 Autocorrelation function and the corresponding relaxation time spectrum of sodium poly(styrene sulfonate) in 3.7 mol/L NaCl. *Reproduced from Sedlák M: What can be seen by static and dynamic light scattering in polyelectrolyte solutions and mixtures?* Langmuir *15(12):4045–4051, 1999.*

performed with commercial instruments and requires different scattering vectors (q), that is, multiple-angle analysis.

Once these limitations regarding particle geometry have been resolved, the conformational variations of polymers and the denaturation or aggregation of proteins can be studied using DLS (Elofsson et al., 1996). For these systems, the autocorrelation functions and relaxation time spectra are typically analyzed, preferentially in consideration of the size distribution analyses (Fang et al., 1991). Fig. 4.5 shows the curve generated by the autocorrelation function and the corresponding spectrum of relaxation times obtained by DLS of an aqueous solution of polyelectrolyte sodium poly(styrene sulfonate) in 3.7 mol/L NaCl (Sedlák, 1999). Two different diffusion modes can be observed: a short-time mode, corresponding to the Na^+ and Cl^- ions, and a slower mode, corresponding to the polyion. This demonstrates that it is possible to separate different diffusive processes in the same system by using DLS.

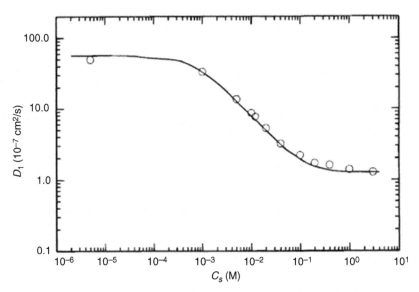

FIGURE 4.6 Diffusion coefficient of sodium poly(styrene sulfonate) as a function of the NaCl concentration added to the aqueous solution. *Reproduced from Sedlák M: What can be seen by static and dynamic light scattering in polyelectrolyte solutions and mixtures?* Langmuir 15(12):4045–4051, 1999.

Fig. 4.6 shows that the diffusion coefficient of the negatively charged polymer depends on the concentration of the low-molar-weight electrolyte added to the solution due to repulsion shielding among the charged groups in the polyion chain. This effect results in the folding of the polymer chains, which is reflected in a reduced diffusion coefficient.

The diameter/radius calculated using the DLS technique refers to the diameter of a sphere that moves at the same velocity of the scattering center, that is, the size obtained from the calculation is the diameter of a sphere with the same translational diffusion coefficient of the particle. If a polymer is dispersed in a θ solvent (i.e., a rigid sphere conformation), the DLS measurements will show a size smaller than that of a polymer dispersed in a good solvent. If the scattering center is a polymer in a *random coil* conformation, the calculated size will not have the turning radius calculated in static light scattering but will have that of a sphere moving at the same velocity as the polymer chain in that particular solvent. Most current DLS instruments allow for in situ temperature variations, which enables the study of conformational transitions and aggregations in different systems as a function of temperature (Elofsson et al., 1996). Considering that limitations are mainly associated with particle geometry, the DLS technique provides one of the most practical and fastest ways to study size distributions in monodispersed and polydispersed systems and the kinetics of size evolution in different types of materials.

List of abbreviations and symbols

d	Diameter of the particle
D	Translational diffusion coefficient
DLS	Dynamic light scattering
$G(t)$	Autocorrelation function
$I(t_0)$	Number of photons that reach the detector at the time t_0
$I(t_0 + t)$	Number of photons that reach the detector at the time $t_0 + t$
I_0	Intensity of the incident radiation
k	Boltzmann constant
K_f	Final scattering vector
K_i	Initial scattering vector
L	Length
n	Refractive index
PDI	Polydispersity index
q	Resulting scattering vector
r	Distance between the light source and the scattering center
R_H	Hydrodynamic radius
T	Kelvin temperature
TEM	Transmission electron microscopy
η	Viscosity of the media
λ	Wavelength
θ	Scattering angle
σ	Standard deviation
τ	Relaxation time

References

Badaire S, et al: In-situ measurements of nanotube dimensions in suspensions by depolarized dynamic light scattering, *Langmuir* 20(24):10367–10370, 2004.

Berne JB, Pecora R: *Dynamic light scattering with applications to chemistry, biology, and physics*, 2ª ed., New York, 2000, Dover Publications.

Bohren CF, Huffman D: *Absorption and scattering of light by small particles*, New York, 1983, John Wiley.

Borissevitch IE, Parra GG, Zagidullin VE, Lukashev EP, Knox PP, Paschenko VZ, Rubin AB: Cooperative effects in CdSe/ZnS-PEGOH quantum dot luminescence quenching by a water soluble porphyrin, *J Luminescence* 134:83–87, 2013.

Elofsson UM, Dejmek P, Paulsson MA: Heat-induced aggregation of β-lactoglobulin studied by dynamic light scattering, *Int Dairy J* 6(4):343–357, 1996.

Fang L, Brown W, Konak C: Dynamic light scattering study of the sol–gel transition, *Macromolecules* 24(26):6839–6842, 1991.

Gao J, Ndong RS, Shiflett MB, Wagner NJ: Creating nanoparticle stability in ionic liquid [C₄mim][BF₄] by inducing solvation layering, *ACS Nano* 9:3243–3253, 2015.

Guidelli EJ, Ramos AP, Zaniquelli MED, Baffa O: Green synthesis of colloidal silver nanoparticles using natural rubber latex extracted from *Hevea brasiliensis*, *Spectrochim Acta A Mol Biomol Spectrosc* 82:140–145, 2011.

Guidelli EJ, Ramos AP, Zaniquelli MED, Nicolucci P, Baffa O: Synthesis and characterization of silver/alanine nanocomposites for radiation detection in medical applications: the influence of particle size on the detection properties, *Nanoscale* 4:2884–2993, 2012.

Jana NR, Earhart C, Ying JY: Synthesis of water-soluble and functionalized nanoparticles by silica coating, *Chem Mater* 19(21):5074–5082, 2007.

Koppel DE: Analysis of macromolecular polydispersity in intensity correlation spectroscopy: the method of cumulants, *J Chem Phys* 57(11):4814–4820, 1972.

Malvern Instruments: *Zeta sizer nano series: user manual*, Worcestershire, 2003, Malvern Instruments Ltd.

Martin JRS, et al: Structure of multiresponsive brush-decorated nanoparticles: a combined electrokinetic, DLS, and SANS study, *Langmuir* 31:4779–4790, 2015.

Provder T: *Particle size distribution: assessment and characterization*, Chicago, 1987, American Chemical Society.

Radeva T: *Physical chemistry of polyelectrolytes*, New York, 2001, Marcel Dekker.

Schmitz KS: *Dynamic light scattering by macromolecules*, San Diego, 1990, Academic Press.

Sedlák M: What can be seen by static and dynamic light scattering in polyelectrolyte solutions and mixtures? *Langmuir* 15(12):4045–4051, 1999.

Shaw DJ: *Introdução à Química dos Colóides e de Superfícies*, São Paulo, 1975, Edgard Blücher Ltda.

Tarasov A, Goerts V, Goodlin E, Nirschl H: Hydrolytic stages of titania nanoparticles formation jointly studied by SAXS, DLS, and TEM, *J Phys Chem C* 117:12800–12805, 2013.

Van Mergen W, Underwood SM: Dynamic-light-scattering study of glasses of hard colloidal spheres, *Phys Rev E* 47(1):248–261, 1993.

Yokoyama M, Satoh A, Sakurai Y, Okano T, Matsumura Y, Karizoe T, Kataoka K: Incorporation of water-insoluble anticancer drug into polymeric micelles and control of their particle size, *J Control Release* 55:219–229, 1998.

Zaru M, Mourtas S, Klepetsanis P, Fadda AM, Antimisiaris SG: Liposomes for drug delivery to the lungs by nebulization, *Eur J Pharm Biopharm* 67:655–666, 2007.

5

X-Ray Diffraction and Scattering by Nanomaterials

Diego G. Lamas*, Mario de Oliveira Neto**, Guinther Kellermann[†],
Aldo F. Craievich[‡]

*CONICET/SCHOOL OF SCIENCE AND TECHNOLOGY, NATIONAL UNIVERSITY OF SAN
MARTÍN, SAN MARTÍN, ARGENTINA; **INSTITUTE OF BIOSCIENCES, UNIVERSITY OF SÃO
PAULO STATE, BOTUCATU - SP, BRAZIL; [†]DEPARTMENT OF PHYSICS, FEDERAL UNIVERSITY
OF PARANÁ, CURITIBA - PR, BRAZIL; [‡]INSTITUTE OF PHYSICS, UNIVERSITY OF SÃO PAULO,
SÃO PAULO - SP, BRAZIL

CHAPTER OUTLINE

1 X-Ray Diffraction Applied to the Study of Nanocrystalline Powders

1.1 X-Ray Diffraction

1.1.1 Introduction

The X-ray diffraction (XRD) technique was first proposed in 1912 by the German physicist Max von Laue (von Laue, 1915). The nature of X-rays, discovered by Wilhelm Röntgen in late 1895, was unknown at the time; knowledge about the structure of matter was also rudimentary. Thus, von Laue's experiment was revolutionary. Von Laue aided by Walter Friedrich and Paul Knipping directed an X-ray beam toward a copper sulfate crystal, as schematically shown in Fig. 5.1. By analyzing the observed diffraction pattern, they simultaneously demonstrated that X-rays are electromagnetic waves whose wavelengths are very short compared with those of visible light and that crystals consist of ordered sets of atoms distributed periodically in space, with characteristic distances of the same order. These important discoveries granted von Laue the 1914 Nobel Prize in Physics.

In 1913, William Henry Bragg and William Lawrence Bragg (father and son), recipients of the 1915 Nobel Prize in Physics, determined the first crystal structures using X-rays. They studied and characterized the atomic order of sodium chloride and other simple compounds. A few years later, researchers began to apply the technique to determine the

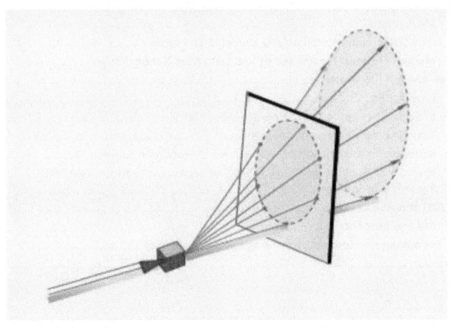

FIGURE 5.1 Schematic view of the experiment conducted by Max von Laue in 1912, which resulted in the development of X-ray diffraction and modern crystallography.

crystal structures of more complex inorganic compounds, originating the scientific area now known as "Crystallography."

The importance of XRD was immediately evident, clearing the path for determining the atomic or molecular structures of all types of materials, which is a prerequisite for understanding their properties. Currently, this technique encompasses a wide range of applications, from relatively simple—but very important—studies in materials science to the structural determinations of complex biological macromolecules, including proteins, nucleic acids, and ribosomes.

1.1.2 Kinematic Theory

The XRD phenomenon can be well understood in terms of the kinematic theory (Giacovazzo et al., 2011). The most important aspects of this theory are briefly discussed in the subsequent text.

XRD originates from the interaction of photons with electrons in the material by means of an elastic (no energy loss) and coherent (well-defined phase relationship between the incident wave and the scattered wave) scattering process. The result of the interaction of an electromagnetic wave with an electron is known in electromagnetism as "Thomson scattering". The scattering intensity, for a non-polarized wave, measured by a detector at a distance r from the electron, is obtained from the following expression Cullity (1956):

$$I(2\theta) = I_0 \frac{e^4}{r^2 m^2 c^4} \left(\frac{1 + \cos^2(2\theta)}{2} \right) \tag{5.1}$$

where 2θ is the angle between the directions of the incident beam and of the scattered beam (scattering angle), m and e are the mass and charge of the electron, respectively, c is the speed of light in vacuum, and I_0 is the intensity of the incident X-ray beam. The term in parentheses is known as the "polarization factor."

For an atom with atomic number Z (Z being the number of electrons in the atom), one might think that the photons would be independently scattered Z times, and the intensity produced by the atom $I_{at}(2\theta)$ would thus be Z times the value of $I(2\theta)$, based on Eq. (5.1). This assumption does not hold true, however, as one must take into account the fact that the electron density is distributed in a finite region around the nucleus. This distribution produces interference effects that influence the amplitude of the scattered wave. As a reminder, the amplitude of the scattered wave $A(2\theta)$ is related to its intensity $I(2\theta)$ by the expression $I(2\theta) = |A(2\theta)|^2$. The ratio between the amplitudes of waves scattered by an atom and by an electron, $f(2\theta)$, is the so-called "atomic form factor" (also known as the "atomic scattering factor"). Fig. 5.2 shows the atomic scattering factor of copper as a function of $\sin\theta/\lambda$. In this case, $f(0) = Z_{Cu} = 29$. In the general case, only when the scattering angle $2\theta = 0°$ does the total amplitude of the scattered wave equals Z times the amplitude scattered by an electron.

In order to determine the total scattering intensity produced by a cluster of atoms, the phase differences associated to the scattering by each atom, due to their different positions, should be accounted for. As the spatial order of the system increases, increasing constructive and destructive interference effects are observed, which, in the case of crystal

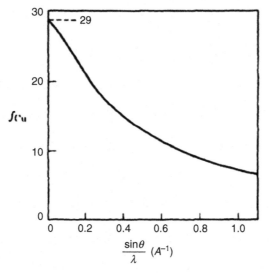

FIGURE 5.2 Atomic scattering factor of copper.

diffraction, generate peaks in well-defined directions. For instance, the diffraction patterns of an amorphous solid (whose atomic structure is close to that of a liquid) and a crystalline solid are compared in Fig. 5.3. The latter reveals the presence of well-defined peaks at particular scattering angles, whereas the former shows an intensity maximum that spans several degrees along 2θ.

In the particular case of a perfect crystal—that is, a material with periodically spaced atoms—the scattering intensity is significant only along a few well-defined directions that comply with Bragg's law:

$$\lambda = 2d_{hkl} \sin\theta_{hkl} \tag{5.2}$$

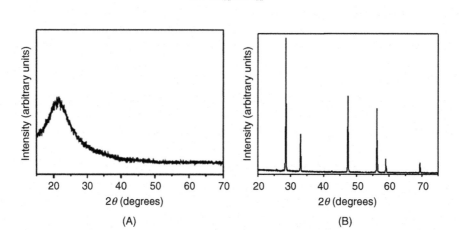

FIGURE 5.3 Qualitative plots of X-ray scattering intensities as functions of the scattering angle produced by (A) an amorphous solid or a liquid and (B) a crystalline solid.

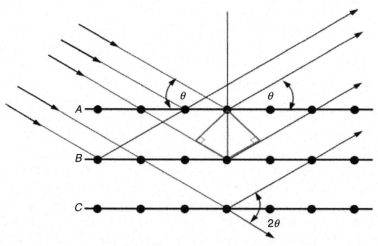

FIGURE 5.4 Scheme for the deduction of Bragg's law. For the family of crystallographic planes drawn, constructive interference occurs when the optical path difference of the different beams (marked in red) is equal to an integer number of wavelengths (*nλ*).

where *λ* is the X-ray wavelength, d_{hkl} is the distance between successive crystallographic planes of a given family of parallel planes defined by the Miller indices *h k l*, and *θ* is the angle between the direction of the incident beam and the crystallographic plane, which is equal to half of the scattering angle *2θ*. Bragg's law can be deduced from the scheme in Fig. 5.4, assuming that a crystal with atoms or molecules regularly arranged in three dimensions behaves as a diffraction grating for X-rays. Therefore, constructive interference occurs when the optical path difference of the different beams is equal to an integer multiple of wavelength *λ*, that is, *nλ*.

The ratio between the amplitude of the wave scattered by *n* atoms belonging to each structural unit cell unit cell atoms and the amplitude scattered by a free electron is a complex magnitude known as the F_{hkl} structure factor. This factor is the sum of the scattering factors of each unit cell atom, f_j, multiplied by their respective phase factors (Giacovazzo et al., 2011):

$$F_{hkl} = \sum_{j=1}^{n} f_j\, e^{2\pi i (h x_j + k y_j + l z_j)} \tag{5.3}$$

The sum in Eq. (5.3) includes all atoms of the unit cell with fractional coordinates x_j, y_j, and z_j. The integers *h*, *k*, and *l* are the Miller indices associated with crystallographic planes *hkl* corresponding to the different diffraction peaks. This expression takes into account the constructive and destructive interference due to the different optical paths of the rays associated with the waves scattered by each unit cell atom.

In turn, it can be easily demonstrated that, for conditions under which Bragg's law is satisfied, the total amplitude of the X-ray wave scattered by *N* unit cells of a crystal is *N* times the amplitude of the wave scattered by a set of *n* unit cell atoms (i.e., all partial waves are "in phase"). Under these conditions, the total scattering intensity is proportional to the product of the structure factor and its complex conjugate or to the square of the structure factor modulus (Giacovazzo et al., 2011). The structure factor can be defined with more accuracy to include the effects of atom vibrations. In addition, other factors must also be considered when calculating the integrated intensities of Bragg peaks for the crystalline powder method (described in more detail in the following sections).

Until now, this discussion has focused only on cases of a single crystal (monocrystal) whose crystallographic characteristics can be analyzed from X-ray scattering intensity measurements in specific directions and for adequate sample orientations, as illustrated in Fig. 5.1. However, it is important to note that it is not possible to measure the phase factors of scattering amplitudes or structure factors. From the intensity of Bragg peaks, I_{hkl}, only the moduli of the structure factors can be calculated, $|F_{hkl}| \propto (I_{hkl})^{1/2}$. Thus, valuable crystallographic information, such as the set of phase factors F_{hkl}, is not experimentally accessible. In crystallography, this drawback is known as the "phase problem" (Giacovazzo et al., 2011).

Dynamical theory, first developed by Paul P. Ewald, constitutes a more complete and accurate theoretical description of XRD than that presented here (Giacovazzo et al., 2011). However, as it will be described in following subsections, the kinematical theory is sufficient for the detailed analysis of X-ray diffraction patterns produced by crystalline powders.

1.2 Powder Diffraction Method

1.2.1 Fundamentals

XRD applied to crystalline powders—X-ray powder diffraction (XPD)—is a variation of the XRD technique for single crystals, in which the sample is a polycrystalline material with a large number of small, randomly oriented crystals (Klug and Alexander, 1974). This method is also applied to powders formed by crystallites with preferred orientations. Typically, powders of polycrystalline materials with a crystallite size of a few micrometers are studied. In this case, it is assumed that for crystallographic planes with any interplanar distance d_{hkl}, there is always a significant fraction of properly oriented crystals that satisfy Bragg's law, and, therefore, all reflections that meet the $|F_{hkl}| \neq 0$ criterion will be observed experimentally.

Note that XPD experiments do not yield diffracted beams that follow well-defined single directions. Instead, there are diffracted beams for a set of directions that make up the so-called "Bragg diffraction cones," as shown in Fig. 5.5. This phenomenon occurs because for an ideal powder formed by a large number of randomly oriented crystallites, there will always be a number of crystallites whose crystallographic planes form an angle θ relative to the incident beam, which diffract the X-ray beam along the different possible directions around the incident beam. Therefore, for each interplanar distance d_{hkl} in a given crystalline powder, a diffraction cone with an opening of 4θ is obtained, and, thus, the diffraction intensity corresponding to all of these cones can be recorded simultaneously.

Conventional laboratory diffractometers generally uses a step-scanning procedure and an X-ray detector with small sensitive area, or, alternatively, a one-dimensional position sensitive detector, in both cases allowing one to only measure the intensities associated with a very small fraction of the diffraction cones, thus resulting in a loss of most of the

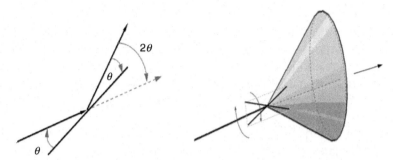

FIGURE 5.5 The origin of the diffraction cones observed using the XPD method. *XPD,* X-ray powder diffraction.

diffracted beam. However, two-dimensional detectors allow for the simultaneous collection of diffraction intensities corresponding to entire diffraction cones. Two-dimensional plane position-sensitive detectors are generally used when the incident X-ray beam is generated by a synchrotron light source and has a high photon energy E, that is, short wavelength $\lambda [\lambda (\text{Å}) = 12.39/E(\text{keV})]$. In this case, all Bragg angles are smaller than those observed for lower energies, thus enabling one to measure intensities associated with a greater number of cones. This type of measurement is quicker, often exhibits better photon-counting statistics, and is frequently used to study in situ phase transitions in materials in real time.

The kinematic theory applied to the XPD method allows one to determine the intensity of Bragg peaks $I_{hkl}(2\theta)$, with the Miller indices $(h\,k\,l)$, as a function of half of the scattering angle θ, according to the following equation (Giacovazzo et al., 2011):

$$I_{hkl}(2\theta) = A\,|\,F_{hkl}\,|^2\, m_{hkl} L(\theta) P(\theta) \tag{5.4}$$

where A is a constant, independent of the scattering angle 2θ, F_{hkl} is the structure factor including the effects of atomic vibrations (Debye–Waller factor), m_{hkl} is the multiplicity of crystallographic planes associated with the Bragg peaks with index hkl, $L(\theta)$ is the Lorentz factor, and $P(\theta)$ is a polarization factor. The Lorentz factor depends on the experimental setup used; in the case of a randomly oriented powder with symmetric scanning (as is the general case for conventional diffractometers, as discussed later), the Lorentz factor is expressed as $L(\theta) = \cos(\theta)/\sin^2(2\theta)$. In experiments with unpolarized incident beams, the polarization factor corresponds to the Thomson scattering [Eq. (5.1)] that occurs when using a conventional X-ray tube, without the use of monochromatic crystals. When using classical X-ray sources with crystal monochromators or synchrotron sources, the beam is polarized and then the polarization factor is different from that defined by Eq. (5.1). In the case of synchrotron sources, the radiation is linearly polarized in the plane of the electron orbit, and, for usual experimental setups, the polarization factor is independent of the scattering angle, $P(\theta) = 1$.

Important applications of the XPD method include (1) qualitative analysis of the crystalline phases present in a given polyphasic material, which are identified by comparing the experimental diffractograms with standards found in crystallographic databases of known compounds, and (2) quantitative determination of the volume or weight fractions of the various phases. XPD is also commonly applied to the "refinement" of atomic structures in the phases present using the Rietveld method (Young, 1995). The Rietveld method allows one to obtain accurate crystallographic parameter values; however, the refinement should always be initiated using approximate values. Currently, it is also possible to solve an unknown crystal structure exclusively from XPD data by employing analysis strategies similar to those used in single-crystal diffraction. This analytical procedure is complex and generally requires high-quality data, typically obtained through the use of a synchrotron source.

1.2.2 Experimental Settings

For the structural characterization of crystalline powders, classical diffractometers installed in X-ray laboratories use the focusing "Bragg-Brentano geometry" (Klug and Alexander, 1974). As illustrated in Fig. 5.6, the sample has a flat surface, perpendicular to the plane of the drawing, and the divergence of the incident beam is defined by a slit located between the F1 source and the sample. The sample always maintains a symmetric orientation with respect to the incident and diffracted beams (via a rotation of θ); thus, the beams diffracted during the θ-2θ scan converge at the positions of the resolution aperture of the detector, F2. In the setup shown in Fig. 5.6, a crystal monochromator (graphite or other single crystals) is placed in the path of the diffracted beam in order to remove unwanted radiation components (such as $K\beta$ radiation and the continuous portion of the emission spectrum).

The convergence of the diffracted beam that occurs in Bragg–Brentano geometry would be perfect if the sample had a curved surface following the so-called Rowland or focusing circle, which contains points F1 and F2 and the point defined by the sample's axis of rotation. This condition is difficult to achieve in practice, especially because the Rowland circle changes its radius during the θ-2θ scan. In classical XRD setups with Bragg-Brentano geometry, two configurations can be used: (1) a fixed incident beam with simultaneous rotation of the sample, θ, and of the detector, 2θ; or (2) a fixed (usually horizontal) sample, with the X-ray tube (incident beam) and detector (diffracted beam) rotating θ-θ

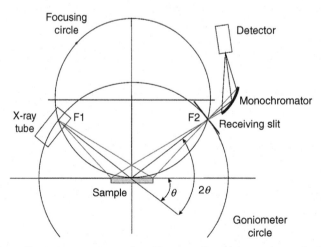

FIGURE 5.6 The basic Bragg–Brentano geometry employed in conventional laboratory powder diffractometers.

in opposite directions. In both setups, the sample always remains tangent to the Rowland circle, as a simple way to achieve convergence of the diffracted beams. The nonideal geometry of this configuration (i.e., because the sample is flat) induces a small instrumental contribution to the width of the Bragg peaks.

The main experimental difficulty encountered in using the Bragg–Brentano configuration is the occurrence of a shift in Bragg peaks when the surface of the sample does not exactly coincide with the diffractometer's axis of rotation and/or when the sample has high transparency (Klug and Alexander, 1974). More accurate results can be obtained with the "parallel incident beam" configuration, which does not induce position shifting of Bragg peaks due to sample eccentricity. This configuration is most often used in XPD stations associated with synchrotron light sources, although it is also utilized in some laboratory equipment.

Although a detailed description of the features of XPD beam lines associated with synchrotron X-ray sources is beyond the scope of this chapter, their main characteristics are listed below.

- High intensity of the incident X-ray beam compared to those produced by conventional sources.
- The possibility of using an incident beam with a very small section, thereby allowing for an accurate selection of specific areas of the sample.
- The possibility of high angular resolution in diffraction or scattering experiments.
- Emission of white radiation, which allows for an arbitrary selection of the incident beam wavelength. This can be applied to differentiate elements with close atomic numbers based on variations in the scattering factor in the vicinity of the absorption edge of some chemical elements present in the sample. This phenomenon is known as anomalous or resonant scattering. Section 3.2 describes an application of anomalous scattering in small-angle X-ray scattering (SAXS) experiments.
- The possibility of using a highly monochromatic incident X-ray beam (through the use of high-resolution single-crystal monochromators), unlike conventional laboratory apparatuses in which the diffraction patterns usually exhibit the $K\alpha_1$–$K\alpha_2$ doublet. Note that conventional XRD apparatuses can employ monochromators that allow for the selection of only $K\alpha_1$ radiation.
- The possibility of using experimental settings with parallel incident beam geometry.

1.3 X-ray Power Diffraction Applied to Nanomaterials

The XPD technique can be used to study nanomaterials with no significant difference from the case of a microcrystalline material. However, it should be noted that there is considerable broadening of the Bragg peaks, as the assumption of an ideal, perfect, and infinite crystal is not satisfied in nanocrystals. This aspect will be addressed in detail in the next section, which will show the inverse correlation between average crystallite size and peak width.

Therefore, the XPD technique applied to nanomaterials provides important information on the external morphology of sample crystallites, whereas in the case of microcrystalline materials, only internal crystallographic information is obtained. In other applications, for example, in the qualitative and quantitative analyses of crystalline phases or the refinement of atomic structure, among others, the procedures are similar to those employed in the case of microcrystalline powders.

1.3.1 Phase Identification by XPD: Application to Nanomaterials

Phase identification is one of the most common applications of XPD, in both microstructured materials and nanomaterials. However, the broadening of Bragg peaks produced in nanocrystalline systems, with consequent height reduction, can hinder the observation of very close or low-intensity peaks. In such cases, it is necessary to collect photon-counting data with good statistical quality and to use mathematical tools in order to separate the contributions corresponding to each peak or to model the entire diffractogram, as in the Rietveld method, for example. From this analysis, one can obtain crystallographic information in a manner similar to that employed for microcrystalline materials.

It is important to take some basic precautions when attempting to identify the phases present in nanomaterials. Some researchers prefer to grow the crystals using thermal treatments at relatively high temperatures to reduce peak widths, allowing the separation of close peaks. This procedure is not recommended, as it assumes that larger crystals have the same phases as the original nanocrystals, which is usually not the case. In nanomaterials, the surface free energy and the interface energy cause significant changes in the thermodynamic properties (Abdala, 2010; Abdala et al., 2010; Acuña, 2012; Acuña et al., 2011; Fábregas, 2008; Fábregas et al., 2008; Garvie, 1965; Lamas, 1999; Lamas et al., 2005, 2006). Hence, phase diagrams of nanostructured systems can be very different from catalogued phase diagrams, which were built considering only the bulk Gibbs free energy. The melting temperature, solubility range, and relative stability of phases may change substantially, for instance. However, the new variable to consider in phase diagrams—crystallite size—has not been extensively considered in the literature nor comprehensively explored from a technological point of view. Furthermore, it is important to always remember that the nanomaterial of interest should be analyzed without applying any treatment that may induce modifications.

A long-known, classic example of retention of metastable phases in nanomaterials is the case of the tetragonal phase in ZrO_2 (zirconia)-based nanoceramics. Pure zirconia exhibits three phases depending on the temperature, with structures based on monoclinic, tetragonal, and cubic Bravais lattices (Juárez et al., 1999). The monoclinic phase is stable at room temperature, but is of little technological interest. By increasing the temperature to 1170°C, a martensitic transformation to the tetragonal phase is observed, while at 2370°C, another martensitic transformation yields the cubic phase. The cubic phase has the structure of fluorite (CaF_2), whereas the tetragonal and monoclinic phases have similar unit cells, distorted only in relation to the cubic cell. The high-temperature phases have excellent electrical and mechanical properties, while the monoclinic phase is not technologically interesting.

While the monoclinic phase is stable at room temperature in the absence of additives, in nanocrystalline powders, the tetragonal phase is retained when the average crystallite size is very small, below a certain critical size of approximately 20 nm. Previous studies have reported critical sizes between 20 and 30 nm (Abdala, 2010; Fábregas, 2008; Garvie, 1965). Some researchers claim that it is also possible to retain the cubic phase at room temperature without additives if the crystallite size is very small—less than 10 nm (Chatterjee et al., 1994; Clearfield, 1964; Mazdiyasni et al., 1966; Roy and Ghose, 2000; Štefanić et al., 1997). However, the cubic phase could not be clearly distinguished from the tetragonal phase in the aforementioned studies, as phase analysis using XRD was carried out under relatively unfavorable conditions using classical diffractometers associated to conventional X-ray sources (instead of synchrotron sources); for this reason, the retention of the cubic phase could not be confirmed (Lamas et al., 2006).

Another interesting example refers to nanocrystalline ZrO_2–Sc_2O_3 solid solutions. This system is of great technological importance due to its possible application as solid electrolyte in solid oxide fuel cells (SOFCs), based on its high ionic conductivity at elevated temperatures compared to those of other ZrO_2-based materials. However, at temperatures below 550–600°C, the equilibrium phases β, γ, and δ, with a rhombohedral structure, exhibit low ionic conductivity. This trend is also observed for the aforementioned monoclinic phase, which exhibits technologically undesirable properties. At high temperatures, the rhombohedral phases transform into the cubic phase, which has excellent electrical properties (Abdala, 2010).

XPD measurements conducted at room or high temperature in the Brazilian Synchrotron Light Laboratory [Laboratório Nacional de Luz Síncrotron (LNLS)] using synchrotron radiation were employed to determine the temperature versus composition phase diagram of nanocrystalline solid solutions with a crystallite size of approximately 25 nm (Abdala et al., 2009). The phase diagram obtained clearly differs from the known equilibrium phase diagram—which was determined for micrometric or larger grain sizes—as the reduction of crystallite size avoids the rhombohedral phases with low ionic conductivity (Abdala, 2010; Abdala et al., 2009).

Fig. 5.7 shows a comparison between two phase diagrams. The phase diagram associated with nanocrystals is qualitatively similar to those of other ZrO_2-based systems, which exhibit t' and t'' tetragonal forms and a cubic phase, but without rhombohedral phases. For a better understanding of the influence of crystallite size on the retention of metastable phases, a detailed crystallographic study was conducted on materials containing 10-14mol% of Sc_2O_3 with crystallite sizes between 10 and 100 nm. The studies revealed the existence of a critical crystallite size—approximately 35 nm—above which the rhombohedral equilibrium phases with low ionic conductivity begin to appear (Abdala et al., 2010, 2012).

Recently, it was found that it is also possible to avoid rhombohedral phases in submicrometric grains of dense ceramics due to the combined effect of small grain size and intergranular stress (Abdala, 2010). This result is of great importance for technological applications of dense ceramics. Impedance spectroscopy studies have shown that these dense ceramics, of small grain size, exhibit excellent ionic conductivity in the temperature range of interest by avoiding the rhombohedral phases.

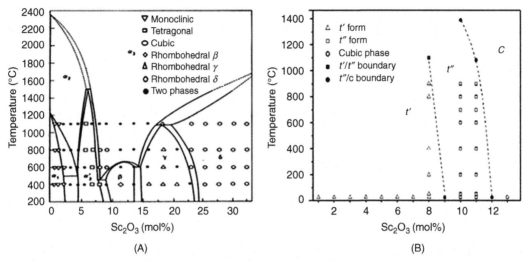

FIGURE 5.7 Phase diagrams of the ZrO₂–Sc₂O₃ system. (A) Equilibrium diagram of microcrystalline material. (B) Phase diagram of a material consisting of 25-nm nanocrystals (Abdala et al., 2009). *Part A based on Ruh R, Garrett HJ, Domagala RF, Patel VA: The system zirconia–scandia. J Am Ceram Soc 60:399–403, 1977.*

1.3.2 Methods for Studying Crystallite Size and Microstrains by Analysis of X-Ray Diffraction Peak Profiles

The analysis of XRD peak profiles to study the microstructural properties of materials is almost as old as XRD itself. Ninety years ago, P. Scherrer found that the width of the diffraction peaks varies inversely with the sample's average crystallite size, D, and described the relationship as follows (Langford and Wilson, 1978):

$$\beta_S = \frac{K\lambda}{D\cos\theta} \tag{5.5}$$

where λ is the wavelength of the incident radiation, θ is the Bragg angle, β is some measure of the peak width, and K is a constant close to 1, known as the Scherrer constant, which depends on the shape of the crystallite and the definitions used for β_S and D. The width measurement chosen can be the integral breadth (β), the variance (W), or the width at half height (Γ) of the diffraction peak. Eq. (5.5) is known as the "Scherrer equation," wherein β_S (the subscript S corresponds to "size") is independent of the order of reflection.

A.R. Stokes and A.J.C. Wilson found that for the integral breadth of the Bragg peak, defined as the ratio between the peak area and the peak height (i.e., the width of a rectangle with the same height and area as the peak considered), the average size D derived from the Scherrer equation is the volume-weighted average crystallite dimension T in the direction perpendicular to the diffraction planes considered:

$$\langle D \rangle_v = \frac{1}{V}\int T\,dV \tag{5.6}$$

In this case, the Scherrer constant is equal to unity, thus Eq. (5.5) becomes:

$$\beta_S = \frac{\lambda}{\langle D \rangle_v \cos\theta} \tag{5.7}$$

In contrast, in the Warren–Averbach method, D is chosen as the average value of the projected surface in the direction perpendicular to the diffraction planes, herein represented as $\langle D \rangle_a$. It is important to note that $\langle D \rangle_v$ can easily be 100% larger than $\langle D \rangle_a$ (de Keijser et al., 1983). An excellent review of these concepts was reported by Langford and Wilson (1978).

Microstrain-induced peak broadenings were reported in 1925 by Van Arkel, but have since been a cause of debate. Generally, two procedures with two different viewpoints have been employed to address this problem (Langford et al., 1988):

1. A microstrain ($\tilde{e}$) is defined by considering local variations in interplanar spacing (d) as $\tilde{e} = \Delta d / d$, and the integral breadth at the angular range corresponding to ($d + \Delta d$) and ($d - \Delta d$) is calculated, assuming that Bragg's law is valid across the entire range. This approach implies that parts of the sample with spacing ($d + \Delta d$) and ($d - \Delta d$) diffract independently. The following expression is thus obtained "(the subscript D is used to indicate "strains" of the distorted crystal lattice):

$$\beta_D = 4\tilde{e}\tan\theta \tag{5.8}$$

2. The strain component $\varepsilon(n)$ is defined as the average strain between two unit cells separated by n cells in a column perpendicular to the diffraction planes. For a Gaussian distribution of $\varepsilon(n)$, independent of the separation of n cells, that is, $\langle \varepsilon^2(n) \rangle = \langle \varepsilon^2 \rangle$, a Gaussian strain profile is obtained. In this case, the region containing the variation in interplanar spacing is considered to diffract coherently, thus generating the following equation:

$$\beta_D = 2(2\pi)^{1/2}\langle \varepsilon^2 \rangle^{1/2}\tan\theta \tag{5.9}$$

Both viewpoints lead to the same angular dependence, as is evident in Eqs. (5.8) and (5.9). Therefore, the parameters $\tilde{e}$ and $\langle \varepsilon^2 \rangle^{1/2}$ are related by the following expression:

$$\tilde{e} = \frac{1}{2}(2\pi)^{1/2}\langle \varepsilon^2 \rangle^{1/2} \tag{5.10}$$

Eq. (5.10) is valid only if $\varepsilon(n)$ has a Gaussian distribution, that is, if the profile of the X-ray peaks caused by strains is Gaussian. Although frequently used, this distribution is not always applicable.

Various methods may be used to study the combined effect of crystallite size and microstrains on XRD. These methods are described in detail in the subsequent text.

1.3.2.1 WILLIAMSON–HALL PLOTS AND SIMPLE CASES

In 1949, Hall proposed a graphical method to distinguish the effects of strain and size, which was later improved by Hall himself in conjunction with Williamson in 1953 (Williamson and Hall, 1953). These researchers observed that if the peak profile is Lorentzian, the plot of $\beta \cos \theta$ versus $\sin \theta$ is linear. The parameters $\langle D \rangle_v$ and $\tilde{e}$ [Eq. (5.10)] are obtained from the intercept and slope, respectively. The authors also observed that the aforementioned plot shows significant deviations from the linear dependence if a considerable Gaussian component exists. However, in the case of a full Gaussian profile, the plot of $\beta^2 \cos^2 \theta$ versus $\sin^2 \theta$ is linear. Although these simple cases are not observed in practice, plots of $\beta \cos \theta$ versus $\sin \theta$ and $\beta^2 \cos^2 \theta$ versus $\sin^2 \theta$ allow for an initial qualitative assessment of the profile behavior; however, quantitative values of $\langle D \rangle_v$ and $\tilde{e}$ are not attainable. Accurate quantitative values are obtained only if one of the two components, the Gaussian or the Lorentzian, is negligible.

1.3.2.2 THE FOURIER METHOD OF WARREN–AVERBACH

A more general theory that considers the simultaneous effect of crystallite size and microstrains in diffraction peak profiles was first proposed by B.E. Warren and B.L. Averbach in 1950–52 (Klug and Alexander, 1974; Warren, 1990). The analyzed X-ray peak profile $h(x)$—where x refers to the angular variable employed—may be considered a convolution of different effects, some originating from the analyzed sample, yielding an $f(x)$ profile, and others from the instrument, yielding a $g(x)$ profile. Thus, the observed profile $h(x)$ is calculated as the "convolution product" of the functions $f(x)$ and $g(x)$:

$$h(x) = f(x) \times g(x) \tag{5.11}$$

However, it is known that the Fourier transform $H(t)$ of function $h(x)$ is related to $F(t)$ and $G(t)$—the transforms of $f(x)$ and $g(x)$, respectively—by the direct product:

$$H(t) = F(t)G(t) \tag{5.12}$$

Thus, if the instrumental profile $g(x)$ is known a priori (measured with a sample of relatively large, microstrain-free crystallites) when measuring the total profile $h(x)$, the Fourier transform of the profile associated with the sample, $F(t)$, can be calculated, which is the $H(t)/G(t)$ quotient. Finally, the $f(x)$ profile, associated exclusively with the sample, can be obtained from the inverse transform of $F(t)$.

The case in which the $f(x)$ profile is the convolution of the effects due to crystallite size $f_S(x)$ and microstrains $f_D(x)$ will be considered next. Because the Fourier coefficients of the full profile are the product of the coefficients of each of these profiles, it follows that

$$A(n,l) = A^S(n)A^D(n,l) \tag{5.13}$$

where A indicates the Fourier cosine coefficients normalized with the condition $A(0, l) = 1$, n is the harmonic number considered, and l is the reflection order.

As previously mentioned, the size profile coefficient $A^S(n)$ is independent of the reflection order. One example is the simple case of an orthorhombic unit cell, with lattice parameters a_1, a_2, and a_3 and diffraction planes of the form $\{0\ 0\ l\}$. If we consider that the domain responsible for the diffraction, or the crystallite, is formed by columns of unit cells perpendicular to the diffraction planes, it can be demonstrated that (Klug and Alexander, 1974; Warren, 1990)

$$A^S(n) = \frac{N(n)}{N_3} \tag{5.14}$$

$$A^D(n,l) = \langle \cos 2\pi l Z(n) \rangle \tag{5.15}$$

where $N(n)$ is the number of unit cells in the whole sample with a neighboring cell at a distance of n cells in the same column, N_3 is the diffraction domain size in single cell units perpendicular to the $(0\ 0\ l)$ planes, and $Z(n)$ is the difference between the spacings of the unit cells separated into n cells. The $\langle \rangle$ symbol in Eq. (5.15) indicates the average over all pairs of unit cells within a distance of n cells in the same column throughout the entire sample. Defining $\varepsilon(n) = Z(n)/n$, it follows that

$$A(n,l) = \frac{N(n)}{N_3} \langle \cos 2\pi n l \varepsilon(n) \rangle \tag{5.16}$$

Because $A^D(n,\ l)$ depends on l, unlike $A^S(n)$, the effects of size and strain can be separated by taking the logarithm of Eq. (5.16), thus yielding

$$\ln A(n,l) = \ln\left(\frac{N(n)}{N_3}\right) + \ln\langle \cos 2\pi n l \varepsilon(n) \rangle \tag{5.17}$$

If the cosine argument in Eq. (5.17) is small, $\ln A(n,\ l)$ can be expanded into a power series keeping only its first terms:

$$\ln A(n,l) \cong \ln\left(\frac{N(n)}{N_3}\right) - 2\pi^2 n^2 l^2 \langle \varepsilon^2(n) \rangle \tag{5.18}$$

This equation is accurate if $\varepsilon(n)$ has a Gaussian distribution. In the linear plot of $\ln A(n,\ l)$ versus l^2, the intercept and slope values lead to the quantities $A^S(n) = N(n)/N_3$ and $\langle \varepsilon^2 \rangle$, respectively.

To calculate the crystallite size as described earlier, the following procedure can be used (Klug and Alexander, 1974; Warren, 1990): The initial slope of the plot of $A^S(n)$ versus n should be extrapolated to the abscissa axis; this intercept gives the value of N_3. Once this value is determined, the crystallite size is calculated using the expression $\langle D \rangle_a = N_3 a_3$, which is an average value in the direction perpendicular to the diffraction planes, weighted by the surface projected in the same direction. Finally, it is important to note that one can determine the crystallite size distribution from the second derivative of $A^S(n)$ (Klug and Alexander, 1974; Warren, 1990).

1.3.2.3 INTEGRAL BREADTH METHOD

Unlike in the Warren–Averbach method, in this case, hypotheses are made about the sample profile $f(x) = f_S(x) \times f_D(x)$, which is estimated by a Voigt function, that is, the convolution product between a Lorentzian function and a Gaussian function. As shown by Langford (1978), this representation is valid if

$$\underset{(\text{pure Lorentzian})}{2/\pi} \leq \Gamma/\beta \leq \underset{(\text{pure Gaussian})}{2(\ln 2/\pi)^{1/2}} \tag{5.19}$$

where Γ is the width at half height and β is the integral breadth of the Bragg peak.

In the simplest version of this method, developed by de Keijser et al. (1982), the size profile $f_S(x)$ is Lorentzian, and the microstrain profile $f_D(x)$ is Gaussian. Thus, one can analyze both crystallite size and microstrains in a single diffraction peak, estimating the peak shape with a Voigt function and finding their Lorentzian $f_L(x)$ and Gaussian $f_G(x)$ components:

$$f(x) = f_L(x) \times f_G(x) \tag{5.20}$$

Thus, assuming that

$$f_S(x) = f_L(x) \tag{5.21a}$$

$$f_D(x) = f_G(x) \tag{5.21b}$$

and applying Eqs. (5.7) and (5.8), it follows that

$$\langle D \rangle_v = \frac{\lambda}{\beta_L \cos\theta} \tag{5.22a}$$

$$\tilde{e} = \frac{\beta_G}{4\tan\theta} \tag{5.22b}$$

where β_L and β_G are the integral breadths of the Lorentzian and Gaussian components, respectively. de Keijser et al. (1982) also calculated approximate expressions for β_L and β_G as functions of $\phi = \Gamma/\beta$:

$$\frac{\beta_L}{\beta} = 2.0207 - 0.4803\phi - 1.7756\phi^2 \tag{5.23a}$$

$$\frac{\beta_G}{\beta} = 0.6420 + 1.4187\left(\phi - \frac{2}{\pi}\right)^{1/2} - 2.2043\phi + 1.8706\phi^2 \tag{5.23b}$$

If the $f(x)$, $g(x)$, and $h(x)$ functions in Eq. (5.11) are all Voigt functions, this method eliminates the contribution of instrumental effects to peak width by considering the following:

$$h_L(x) = f_L(x) \times g_L(x) \tag{5.24a}$$

$$h_G(x) = f_G(x) \times g_G(x) \tag{5.24b}$$

thus yielding

$$\beta_L^f = \beta_L^h - \beta_L^g \tag{5.25a}$$

$$(\beta_G^f)^2 = (\beta_G^h)^2 - (\beta_G^g)^2 \tag{5.25b}$$

In the most general case, when diffraction peak profiles contain Lorentzian and Gaussian components, the procedure for determining $\langle D \rangle$ and $\tilde{e}$ [for both $f_S(x)$ and $f_D(x)$] requires an analysis of peaks corresponding to two or more orders from the same reflection. This method was proposed by Langford et al. (1988). In this case, the deconvolution process (elimination of instrumental effects) given by Eqs. (5.24a), (5.24b), (5.25a), and (5.25b) remains valid. Assuming that $f(x)$, $f_S(x)$, and $f_D(x)$ are Voigt functions, it follows that

$$\beta_L^f \cos\theta = \beta_L^S + \beta_L^D \sin\theta \tag{5.26a}$$

$$(\beta_G^f)^2 \cos^2\theta = (\beta_G^S)^2 + (\beta_G^D)^2 \sin^2\theta \tag{5.26b}$$

Thus, by plotting $\beta_L^f \cos\theta$ versus $\sin\theta$, β_L^S and β_L^D are obtained, while from the plot of $(\beta_G^f)^2 \cos^2\theta$ versus $\sin^2\theta$, $(\beta_G^S)^2$ and $(\beta_G^D)^2$ are determined. From the integral breadths of the Lorentzian and Gaussian components of $f_S(x)$ and $f_D(x)$, β^S and β^D are calculated using the following expression:

$$\beta = \frac{\beta_G \exp(-k^2)}{1 - \mathrm{erf}(k)} \tag{5.27}$$

where $k = \beta_L/\pi^{1/2}\beta_G$ and erf(x) is the error function. The following approximation can be used instead of Eq. (5.27) (de Keijser et al., 1982):

$$\frac{\beta_G}{\beta} = -\frac{1}{2}k\pi^{1/2} + \frac{1}{2}(\pi k^2 + 4)^{1/2} - 0.234k\exp(-2.176k) \tag{5.28}$$

It is important to note that the β^S and β^D values obtained in this way are the constants for scaling $1/(\cos\theta)$ and $\tan\theta$ in Eqs. (5.7) and (5.8); these values should not be confused with the integral breadth of the previous single-peak model (for distinction, S and D are indicated as subscripts in the former case and as superscripts in the current case). Using Eqs. (5.7) and (5.8), the following equations can be employed to calculate $\langle D \rangle_v$ and $\tilde{e}$ from β^S and β^D:

$$\langle D \rangle_v = \frac{\lambda}{\beta^S} \tag{5.29a}$$

$$\tilde{e} = \frac{\beta^D}{4} \tag{5.29b}$$

Finally, it is important to note that a number of programs for the analysis and fitting of XRD profiles are unable to work with Voigt functions. This is the case, for example, for the well-known Rietveld programs. In contrast, pseudo-Voigt functions (the sum of Gaussian and Lorentzian functions) and Pearson VII functions are typically applied. In this case, a fit is performed with the chosen function, followed by a calculation of the equivalent Lorentzian and Gaussian component breadths. de Keijser et al. (1983) calculated approximate expressions for β_L and β_G according to the characteristic parameters of these functions. To demonstrate this procedure, we will now consider the case of a pseudo-Voigt function:

$$pV(2\theta) = I_0[\eta L(2\theta) + (1-\eta)G(2\theta)] \tag{5.30}$$

where $L(2\theta) = [1 + (2\theta - 2\theta_0)^2/\Gamma^2]^{-1}$ and $G(2\theta) = \exp[-\ln 2(2\theta - 2\theta_0)^2/\Gamma^2]$ are the Lorentzian and Gaussian functions, respectively, $2\theta_0$ is the angular position of the peak under consideration, η is a mixing parameter, and Γ is the half-height peak width. The integral breadth of a pseudo-Voigt function is given by

$$\beta = \frac{\Gamma}{2}\left[\pi\eta + (1-\eta)\left(\frac{\pi}{\ln 2}\right)^{1/2}\right] \tag{5.31}$$

The following approximate expressions can be used to calculate the equivalent β_L and β_G breadths:

$$\frac{\beta_L}{\beta} = 0.017475 + 1.50048\eta - 0.534156\eta^2 \tag{5.32a}$$

$$\frac{\beta_G}{\beta} = 0.184446 + 0.812692(1 - 0.998497\eta)^{1/2} - 0.659603\eta + 0.44542\eta^2 \tag{5.32b}$$

1.4 Rietveld Method and its Application to the Study of Crystallite Size and Microstrains

The Rietveld method allows one to simulate an experimental XRD pattern from a model that includes two main contributions: (1) the atomic structure, based on the average atomic positions in all unit cells, and (2) the various contributions to diffraction peak profiles, represented in terms of analytical functions. Both contributions are important to achieve a good agreement between the simulated and observed diffraction patterns. On a first approximation, the integral intensity of the Bragg reflections and their angular positions depend on the average crystallographic structure. However, peak profiles are related to both the geometry of the apparatus and the sample microstructure. Thus, an analysis of the diffraction peak profiles enables one to obtain relevant information on the structural imperfections of the sample: not only crystallite size but also variations in interplanar distances due to internal strains, nonstoichiometry, twinnings, stacking faults, dislocations, etc.

This section briefly describes the procedures followed when the Rietveld method is applied to study imperfections in materials. The main objective of this type of analysis is to find a tentative physical explanation for Bragg peak broadenings. However, it should be noted that the Rietveld method is most frequently applied in the refinement of crystallographic structural models, and not in the analysis of diffraction peak profiles.

An important feature of the Rietveld method is that it not only takes into account the effects mentioned earlier but also includes parameters for instrumental aspects to correct for off-centering, sample transparency, displacement of the 2θ zero, etc.

The Rietveld method for determining crystallite size and strains employs a procedure similar to that described for the integral breadth. First, the diffraction patterns of the sample and those of the reference sample—with negligible broadening due to crystallite size or imperfections—are fitted with an analytical function for the different peak profiles. The best fits are generally obtained (1) using the pseudo-Voigt or Pearson VII functions to determine the integrated breadths of the Lorentzian and Gaussian components using Eqs. (5.23a) and (5.23b) from $\phi = \Gamma/\beta$ or (2) using the approximations of de Keijser et al. [Eqs. (5.32a) and (5.32b), where the fit is performed using a pseudo-Voigt function] based on the characteristic parameters of that function. Once the β_L^h, β_G^h, β_L^g, and β_G^g integral breadths are determined, the parameters β_L^f and β_G^f are obtained using the deconvolution procedure described by Eqs. (5.24a) and (5.24b).

Finally, the average crystallite size $\langle D \rangle_v$ and microstrain parameter $\tilde{e}$ are determined either (1) by analyzing a single diffraction peak assuming that β_L^f and β_G^f accurately represent the size and strain effects, respectively, or (2) by applying a more general method that considers the various orders of a Bragg reflection, calculating the parameters β_L^S, β_L^D, β_G^S, and β_G^D from Eqs. (5.26a) and (5.26b), followed by the application of Eqs. (5.27) and (5.28) to obtain β^S and β^D.

Several structure refinement programs based on the Rietveld method use the pseudo-Voigt (Gaussian + Lorentzian) function and fit the mixing parameter η of the functions and the half-height width Γ from parameters that model their respective dependencies on 2θ (Young, 1995). For η, a linear relationship is typically used:

$$\eta = N_A + N_B 2\theta \tag{5.33a}$$

To determine Γ, the following dependency is used (Young, 1995):

$$\Gamma^2 = U \tan^2\theta + V \tan\theta + W + \frac{Z}{\cos^2\theta} \tag{5.33b}$$

Thus, the parameters N_A, N_B, U, V, W, and Z are determined by a fitting procedure.

Several powder diffraction programs employ the modified Thompson–Cox–Hastings (TCH) pseudo-Voigt function (Thompson et al., 1987). In this case, a pseudo-Voigt function is used; however, instead of fitting η and Γ, the half-height widths of the Lorentzian and Gaussian components Γ_L and Γ_G, respectively, are fitted directly; subsequently, η and Γ can be calculated using the following equations:

$$\eta = 1.36609q - 0.47719q^2 + 0.1116q^3 \tag{5.34a}$$

$$\Gamma = (\Gamma_G^5 + A\Gamma_G^4\Gamma_L + B\Gamma_G^3\Gamma_L^2 + C\Gamma_G^2\Gamma_L^3 + D\Gamma_G\Gamma_L^4 + \Gamma_L^5)^{0.2} \tag{5.34b}$$

where $q = \Gamma_L/\Gamma$, $A = 2.69269$, $B = 2.42843$, $C = 4.47163$, and $D = 0.07842$.

The widths Γ_L and Γ_G are fitted using the following expressions:

$$\Gamma_L = X\tan\theta + \frac{Y}{\cos\theta} \tag{5.35a}$$

$$\Gamma_G = \left(U\tan^2\theta + V\tan\theta + W + \frac{Z}{\cos^2\theta}\right)^{1/2} \tag{5.35b}$$

In both cases, six parameters are fitted.

It is important to interpret the meaning of the aforementioned parameters. With $\beta_L = (\pi/2)\Gamma_L$ and $\beta_G = (1/2)(\pi/\ln 2)^{1/2}\Gamma_G$, a comparison with Eqs. (5.26a) and (5.26b) yields the following:

$$\beta_L^S = \left(\frac{\pi}{2}\right)Y \tag{5.36a}$$

$$\beta_L^D = \left(\frac{\pi}{2}\right)X \tag{5.36b}$$

$$\beta_G^S = \frac{1}{2}\left(\frac{\pi}{\ln 2}\right)^{1/2}Z^{1/2} \tag{5.36c}$$

$$\beta_G^D = \frac{1}{2}\left(\frac{\pi}{\ln 2}\right)^{1/2}U^{1/2} \tag{5.36d}$$

Before applying these equations, the profiles must be corrected due to instrumental effects. If the instrumental and sample diffraction profiles can be represented by the same function, Eqs. (5.25a) and (5.25b) indicate that instrumental effects can be corrected by a simple subtraction of fitted coefficients. For instance, in neutron diffraction experiments, the instrumental profile is usually approximately Gaussian and may be described using only the parameters U, V, and W from Eq. (5.35b); hence, the simple calculation of $U - U_{ins}$ determines β_G^D. Note that these parameters are independent of the Miller indices $(h\,k\,l)$ of the considered Bragg reflection; thus, Eqs. (5.36a) to (5.36d) are valid for the cases in which size and strain effects are isotropic. Some programs begin by applying isotropic models, introducing corrections that consider the existence of any anisotropy at a later stage.

1.5 Case Study: Nanocrystalline Y_2O_3-Doped ZrO_2 Powders

The case of nanocrystalline Y_2O_3-Doped ZrO_2 powders, synthesized by the gel combustion method using citric acid as fuel (Juárez et al., 2000; Lamas, 1999), exemplifies the methods described in the precedent subsections. A composition ZrO_2–2.8 mol% Y_2O_3 was selected. Ashes obtained after the combustion reaction were calcinated at 600°C for 2 h for slow elimination of carbonaceous residue. In the end, a very light white, foamy powder was obtained, formed by slightly clustered nanoscale crystallites. XPD analysis was conducted using a conventional laboratory Philips PW 3710 X-ray diffractometer (at the Centro de Investigaciones en Sólidos, CONICET-CITEDEF, Villa Martelli, Argentina) using Cu $K\alpha$ radiation.

Fig. 5.8 shows the diffraction pattern of the calcinated powder. The splitting of the $(4\ 0\ 0)_t$ and $(0\ 0\ 4)_t$ Bragg peaks, shown in the inset of Fig. 5.8, indicated the retention of the tetragonal phase. The diffraction peaks shown in Fig. 5.8 were indexed by considering a pseudo–fluorite-type tetragonal unit cell (Juárez et al., 1999; Lamas, 1999). It should be noted that no trace of stable monoclinic phase peaks (usually present in materials with large crystallite size) was observed. It was concluded that the existence of the tetragonal phase and the complete absence of the monoclinic equilibrium phase in the solid solution studied (ZrO_2–2.8 mol% Y_2O_3) at room temperature occur only in nanocrystalline samples.

For an initial calculation of the average crystallite size, the profile of peak $(1\ 1\ 1)_t$ was fitted with a symmetric Pearson VII function. From this fit, the integral breadth was determined and the average crystallite size calculated using Scherrer's equation (5.5) or (5.7), considering the effect of microstrains and instrumental broadening to be negligible. An estimated value of $\langle D \rangle_v = 92(2)$ Å was obtained.

In this case, this estimate is a quantitative value with a certain accuracy because strain-induced broadening is not significant at low angles, given that Bragg peak broadening due

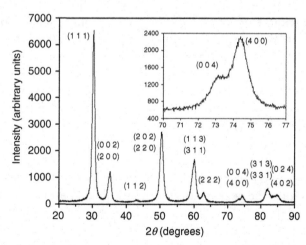

FIGURE 5.8 Diffractogram of Y_2O_3-doped ZrO_2 powder, synthesized by gel combustion, with details regarding the splitting of $(4\ 0\ 0)_t$ and $(0\ 0\ 4)_t$ Bragg peaks (Lamas, 1999).

Table 5.1 Estimated Values of Average Crystallite Size $\langle D \rangle_v$ and Microstrain Parameter $\tilde{e}$ Corresponding to a Y_2O_3-Doped ZrO_2 Nanopowder Doped with Tetragonal Phase, Derived from the Diffraction Pattern Shown in Fig. 5.8

β_L^S (degree)	β_G^S (degree)	β_L^D (degree)	β_G^D (degree)	β^S (degree)	β^D (degree)	$\langle D \rangle_v$ (Å)	$\tilde{e}$ ($\times 10^{-3}$)
0.33	0.50	1.12	—	0.73	1.12	120 (5)	4.9 (2)

Table 5.2 Lattice Parameters and Fractional z-Coordinate of the O^{2-} Ion in the Asymmetric Unit of the Tetragonal Unit Cel Determined by the Rietveld Method Using the Conventional Pseudo-Voigt Function

a (Å)	c (Å)	z (O^{2-})
5.1052 (4)	5.1806 (4)	0.2039 (5)

to this effect is proportional to tan θ. Furthermore, because the diffraction peaks shown in Fig. 5.8 are very broad, it is expected that the instrumental broadening—typically on the order of a few hundredths of a degree, reaching a maximum of approximately 0.1°—will not significantly alter the intrinsic profile. The errors were evaluated from the differences found in the integral breadth of the peaks determined with several fits, without considering the systematic errors mentioned earlier.

Table 5.1 reports the results obtained after applying the Rietveld method with a conventional pseudo-Voigt function [given by Eq. (5.30)] and fitting the parameters for optimization. Through this first analysis, an estimated value of $\langle D \rangle_v = 120(5)$ Å was obtained. It was also possible to obtain an estimate for the parameter $\tilde{e}$. The crystallographic parameters determined by this fitting procedure are shown in Table 5.2 and are in agreement with other data in the literature.

To obtain an initial qualitative understanding, diffraction peak integral breadths were calculated from the parameters fitted with the conventional pseudo-Voigt function using expression (5.31), and a Williamson–Hall plot, $\beta \cos \theta$ versus $\sin \theta$, was constructed (Fig. 5.9). In this case, no instrumental corrections were made in determining the β parameter; however, the effects of $K\alpha_2$ radiation were eliminated. Analysis of Fig. 5.9 shows that although a small curvature is observed, the points approximately lie on a straight line, indicating a predominance of the Lorentzian component of the profile. Furthermore, the high slope value indicates a strong effect arising from strains. Otherwise, a horizontal line is obtained, as predicted by the Scherrer equation. This result agrees with that obtained previously using the $(1\ 1\ 1)_t$ and $(2\ 2\ 2)_t$ reflections, but in this procedure, all Bragg peaks are considered. Other estimates of $\langle D \rangle_v$ and $\tilde{e}$ were calculated from the intercept (b) and line slope (m) values. In these calculations, the peak widths given by Eqs. (5.7) and (5.8) are considered to be additive, which would be exactly valid if both effects generated Lorentzian profiles. Thus, $\langle D \rangle_v = \lambda / b$ and $\tilde{e} = m/4$. The obtained values are reported in Table 5.3.

The above estimations of $\langle D \rangle_v$ and e can be compared with the values determined from the use of the modified TCH pseudo-Voigt function, without considering anisotropy

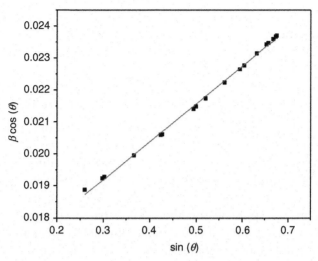

FIGURE 5.9 Williamson–Hall plot, $\beta \cos \theta$ versus $\sin \theta$, obtained from the diffractogram displayed in Fig. 5.8 (Lamas, 1999).

effects. In this case, the instrumental effects determined with an Al_2O_3 powder standard are subtracted. The values of the crystallographic parameters reported in Table 5.4 are consistent with those obtained using a conventional pseudo-Voigt function (Table 5.2). Fitting of the diffraction peak profile again shows that β_G^D is negligible and that the Lorentzian component dominates over the Gaussian. The β_L^S, β_L^D, and β_G^S values calculated by Eqs. (5.36a)–(5.36d) and the results obtained for β^S, $\beta^D \langle D \rangle_v$, and $\tilde{e}$ are reported in Table 5.5.

Table 5.3 Values of $\langle D \rangle_v$ and $\tilde{e}$ Determined From the Williamson–Hall Plot

$\langle D \rangle_v$ (Å)	$\tilde{e}$ ($\times 10^{-3}$)
98 (1)	2.95 (5)

Table 5.4 Lattice Parameters and Fractional z-Coordinate of the O^{2-} Ion in the Asymmetric Unit of the Tetragonal Unit Cell Determined by the Rietveld Method Using the Modified TCH Pseudo-Voigt Function

a (Å)	c (Å)	z (O^{2-})
5.1049 (4)	5.1802 (4)	0.2041 (5)

TCH, Thompson–Cox–Hastings.

Table 5.5 Values of $\langle D \rangle_v$ and $\tilde{e}$ Determined by the Rietveld Method Using the Modified TCH Pseudo-Voigt Function

β_L^S (degree)	β_G^S (degree)	β_L^D (degree)	β_G^D (degree)	β^S (degree)	β^D (degree)	$\langle D \rangle_v$ (Å)	$\tilde{e}$ ($\times 10^{-3}$)
0.51 (1)	0.44 (1)	0.74 (2)	—	0.70 (3)	0.74 (2)	108 (3)	3.2 (1)

TCH, Thompson–Cox–Hastings.

The results for $\langle D \rangle_v$ and $\tilde{e}$ reported in Table 5.5 are more accurate than those calculated previously, as the instrumental effects were subtracted prior to their calculation. Other less complex methods, such as the Williamson–Hall method, allow one to obtain approximate values for the same parameters in a simpler fashion.

1.6 Modern Methods for Analysis of XPD Data from Nanomaterials

Thus far, the classical methods for analyzing Bragg peaks profiles have been presented and discussed, providing a relatively simple way for obtaining information on the average crystallite size and microstrains. These methods can be very useful in improving synthesis processes and in understanding the physicochemical properties of nanostructured materials.

Currently, there are new strategies for determining structural parameters from complete powder diffraction patterns, instead of individual diffraction peak profiles. In particular, the Rietveld method allows one to characterize the crystallographic and microstructural aspects in a single analysis, with subroutines that determine crystallite shape, strain directions, etc.

With the development of computers, the conventional methods discussed in the previous sections have led to more complete procedures known as full-pattern profile fitting (FPPF) and full-pattern profile modeling (FPPM), based on a simultaneous analysis of the entire diffractogram, thus resolving the problems associated with a superposition of diffraction peaks. In FPPM, the diffraction pattern of the entire sample is modeled by convoluting different effects with well-defined physical meaning, such as those due to the distribution of crystallite sizes, dislocations, and twinning. For more information on these methods, the reader is referred to the articles by Scardi and Leoni from the University of Trento, Italy (Mittemeijer and Scardi, 2004; Scardi et al., 2005).

A different, more modern procedure for the study of nanocrystalline materials from XPD patterns—which uses high-energy (i.e., small-wavelength) X-rays—is based on the pair distribution function (PDF). The PDF contains information about the relative distances between atoms and can be utilized to determine the degree of local order in the material, distances between neighboring atoms, coordination numbers, etc. In contrast to the treatment of conventional diffraction data, the PDF method allows one to determine the local atomic order of materials. The results can then be related to X-ray absorption spectroscopy [extended X-ray absorption fine structure (EXAFS)] data, the most commonly used technique for structural analysis of the local atomic order of materials. A detailed description of the PDF method can be found in articles by Petkov from Central Michigan University (Petkov, 2008).

In summary, the XPD technique provides complete and valuable information for the structural characterization of nanomaterials, including information that is not strictly related to crystallography, including crystallite morphology and local atomic order. The recent development of new analytical strategies such as those mentioned earlier clearly indicates that the XPD experimental technique is still evolving.

2 Small-Angle X-Ray Scattering

2.1 Basic Aspects

The forthcoming subsections address the basis of the theory of small-angle X-ray scattering (SAXS), which is developed with more detail in the classical books dedicated to this subject (Guinier and Fournet, 1955; Glatter and Kratky, 1982).

As we have seen, the results of wide-angle X-ray scattering experiments (XRD), such as those presented in Section 1, lead to determinations of high-resolution structural parameters (such as lattice parameters, local strains, atomic positions), while those derived from results of small-angle scattering experiments—to be described in this section and also in Section 3—yield low-resolution (nanoscopic) structural information.

2.1.1 Scattering of X-Rays by Free Electrons

The intensity of elastic X-ray scattering produced by a free electron was determined by Thomson, as specified in Section 1.1.2. For small values of angle 2θ, in the angular range usually used in SAXS experiments (typically $2\theta \leq 6°$), the intensity scattered by an electron per unit of solid angle can be approximated by a constant value (independent of the angle) equal to $I_e(2\theta) = I_0 \cdot r_e^2$, where I_0 is the intensity of the incident X-ray beam (units of power/area) and r_e is the classical electron radius.

In addition to the coherent scattering of X-rays (Section 1.1.2), electrons also produce Compton scattering, which is inelastic in nature. Because the amplitude of the Compton scattering is incoherent (i.e., there is no phase relationship between the incident and scattered waves), the scattering intensity is not modulated by structural correlation effects. However, as the Compton scattering is negligible for small angles, its contribution is ignored when analyzing the results of SAXS experiments.

2.1.2 X-Ray Scattering by a Material Volume with an Arbitrary Structure

We start by considering the general case of a nanostructured material with an arbitrary electron density $\rho(\vec{r})$, and a monochromatic X-ray beam with a wavelength λ hitting the material along a direction defined by a versor $\vec{B}_0$ (Fig. 5.10). To determine the scattering amplitude along any arbitrary direction defined by a versor $\vec{B}_1$, the phase difference between the wave associated with the beam scattered by an element of volume $d\vec{r}$ at the center of mass and another at a position defined by the vector $\vec{r}$ should be calculated.

A difference in optical path Δs corresponding, for example, to rays 1 and 2 in Fig. 5.10A, produces a phase-difference given by

$$\Delta\varphi = \frac{2\pi\,\Delta s}{\lambda} \tag{5.37}$$

For an optical path difference $\Delta s = \overline{AB} + \overline{BC}$ (Fig. 5.10A), the phase difference is

$$\Delta\varphi = \frac{2\pi(\overline{AB} + \overline{BC})}{\lambda} \tag{5.38}$$

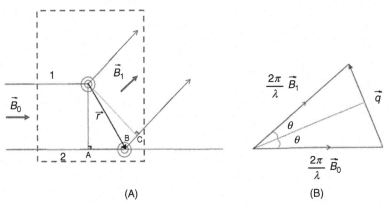

(A)   (B)

FIGURE 5.10 (A) Schematic of the X-ray scattering process for two volume elements, one located at the origin O and the other at a position B defined by vector $\vec{r}$. The directions of the incident and scattered beams are defined by the unit vectors B_0 and B_1, respectively. (B) Definition of the scattering vector $\vec{q}$ as a function of versors $\vec{B}_0$ and $\vec{B}_1$ and of the wavelength λ of the X-ray beam.

It can be seen from Fig. 5.10A that the optical path difference can be written as a function of the $\vec{B}_0$ and $\vec{B}_1$ versors and the position vector $\vec{r}$:

$$\overline{AB} = \vec{r} \cdot \vec{B}_0 \quad \text{and} \quad \overline{BC} = -\vec{r} \cdot \vec{B}_1 \tag{5.39}$$

Consequently, the phase difference can be rewritten as

$$\Delta \varphi = -\frac{2\pi \vec{r} \cdot (\vec{B}_1 - \vec{B}_0)}{\lambda} \tag{5.40}$$

Thus, defining the scattering vector as

$$\vec{q} = \frac{2\pi(\vec{B}_1 - \vec{B}_0)}{\lambda} \tag{5.41}$$

the phase difference becomes

$$\Delta \varphi = -\vec{r} \cdot \vec{q} \tag{5.42}$$

Fig. 5.10B illustrates the relationship between $2\pi \vec{B}_1 / \lambda$, $2\pi \vec{B}_0 / \lambda$, and $\vec{q}$. It is clear that $\vec{q}$ is perpendicular to the bisector of the angle between $\vec{B}_1$ and $-\vec{B}_0$, and its magnitude is given by

$$q = \frac{4\pi \sin \theta}{\lambda} \tag{5.43}$$

For small angles, $q \approx (2\pi/\lambda)(2\theta)$, that is, the scattering vector magnitude, q, is approximately proportional to the scattering angle 2θ.

The amplitude of the wave scattered by the electrons contained in the volume element $d\vec{r}$ is equal to the number of electrons in the volume element $\rho(\vec{r})d\vec{r}$ multiplied by the amplitude of a wave scattered by a single electron, A_e, and the phase factor, $e^{i\Delta\varphi}$:

$$dA(\vec{q}) = A_e \rho(\vec{r}) e^{i\Delta\varphi} \, d\vec{r} \tag{5.44}$$

Finally, the total scattering amplitude is obtained by integrating Eq. (5.44) over the total sample volume, V:

$$A(\vec{q}) = \int_V \rho(\vec{r})e^{i\Delta\varphi}\,d\vec{r} \qquad (5.45)$$

In Eq. (5.45) and subsequent equations, A_e is omitted for simplicity (i.e., it is assumed that $A_e = 1$). Replacing Eq. (5.42) in Eq. (5.45) gives

$$A(\vec{q}) = \int_V \rho(\vec{r})e^{-i\vec{q}\cdot\vec{r}}\,d\vec{r} \qquad (5.46)$$

Thus, the scattering amplitude $A(\vec{q})$—a complex function—is the Fourier transform of the electron density function $\rho(\vec{r})$.

The electron density can be written as $\rho(\vec{r}) = \bar{\rho} + \Delta\rho(\vec{r})$, where $\bar{\rho}$ is the average electron density and $\Delta\rho(\vec{r})$ is the variation in electron density with respect to the average. Thus, the scattering complex amplitude becomes

$$A(\vec{q}) = \int_V \bar{\rho}e^{-i\vec{q}\cdot\vec{r}}\,d\vec{r} + \int_V \Delta\rho(\vec{r})e^{-i\vec{q}\cdot\vec{r}}\,d\vec{r} \qquad (5.47)$$

where the first term of the sum is the Fourier transform of a spatially constant function, $\bar{\rho}$, over a macroscopic volume (total sample volume, V). This result implies that its Fourier transform is zero throughout the entire $\vec{q}$ vectorial space (reciprocal space), except for very small values of $\vec{q}$, which are not accessible in SAXS experiments. Therefore, Eq. (5.47) can be rewritten as

$$A(\vec{q}) = \int_V \Delta\rho(\vec{r})e^{-i\vec{q}\cdot\vec{r}}\,d\vec{r} \qquad (5.48)$$

Eq. (5.48) indicates that the X-ray scattering amplitude is determined by a simple Fourier transform of the function $\Delta\rho(\vec{r})$.

Moreover, the experimental scattering intensity $I(\vec{q})$ is related to the complex amplitude $A(\vec{q})$ by

$$I(\vec{q}) = A(\vec{q})\cdot A(\vec{q})^* = |A(\vec{q})|^2 \qquad (5.49)$$

where $A(\vec{q})^*$ is the complex conjugate of $A(\vec{q})$.

According to the properties of the Fourier transform, the inverse transform of $A(\vec{q})$ is given by

$$\rho(\vec{r}) = \frac{1}{(2\pi)^3}\int A(\vec{q})e^{i\vec{q}\cdot\vec{r}}\,d\vec{q} \qquad (5.50)$$

Eq. (5.50) would allow one to determine the electron density $\rho(\vec{r})$, which is a real and positive function that completely defines the structure of the material. However, to determine $\rho(\vec{r})$ from the integral of Eq. (5.50), the modulus and phase of the complex function $A(\vec{q})$, defined as

$$A(\vec{q}) = |A(\vec{q})|\,e^{i\Delta\varphi} \qquad (5.51)$$

should be known. From the scattering intensity $I(\vec{q})$, which is an experimentally accessible function, the modulus of $A(\vec{q})$ can be determined:

$$|A(\vec{q})| = [I(\vec{q})]^{1/2} \qquad (5.52)$$

This result implies that the scattering amplitude module, $|A(\vec{q})|$, but not its phase, $\Delta\varphi$, can be calculated from the experimental scattering intensity. This problem—known in crystallography as the "phase problem"—was discussed in Section 1.1.2.

2.1.3 X-Ray Scattering by an Atom

The ratio between the X-ray scattering amplitudes produced by an atom and that for an electron (A_e) is called the atomic scattering factor, which was discussed in Section 1.1.2. This factor can be calculated by the following equation:

$$f(\vec{q}) = \int \rho(\vec{r}) e^{-i\vec{q}\cdot\vec{r}} \, d\vec{r} \qquad (5.53)$$

where $\rho(\vec{r})$ is the electron density function of the atom. For atoms in which the electron density depends only on the distance r from their "center of mass" (referring to the electron density), the atomic scattering factors are functions of the modulus of the scattering vector, q.

2.1.4 X-Ray Scattering by a Cluster of Atoms

The scattering amplitude for a cluster of atoms can be calculated from the atomic scattering factor and the position vectors $\vec{r}_j$ of the individual atoms. For this purpose, the scattering factors of n atoms of the scattering object, f_j, are added, with each being multiplied by the corresponding phase factor $e^{-i\vec{q}\cdot\vec{r}_j}$. Thus, the amplitude of X-ray scattering by the whole cluster is given by

$$A_1(\vec{q}) = \sum_{j=1}^{n} f_j \, e^{-i\vec{q}\cdot\vec{r}_j} \qquad (5.54)$$

Eq. (5.54) indicates that the total amplitude of the X-ray beam scattered by a cluster of atoms is the sum of the scattering factors for each atom multiplied by their respective phase factors. As noted in Section 1.1.2, the total scattering amplitude is called the "structure factor." Finally, the X-ray scattering intensity produced by a cluster of n atoms is given by

$$I_1(\vec{q}) = A_1(\vec{q}) A_1(\vec{q})^* = \sum_{j=1}^{n}\sum_{k=1}^{n} f_j f_k \, e^{-i\vec{q}\cdot\vec{r}_{jk}} \qquad (5.55)$$

where $\vec{r}_{jk}$ is the vector from atom j to atom k.

2.1.5 Correlation Function Associated With an Arbitrary Structure

Due to the aforementioned "phase problem," it is not possible to obtain the electron density function $\rho(\vec{r})$ by performing an inverse Fourier transform on $I(\vec{q})$ because only the magnitude, but not the phase, of $A(\vec{q})$ can be determined from the experimental function $I(\vec{q})$. To establish a feasible data analysis procedure, Patterson (1935) proposed using the relationship between the function $I(\vec{q})$ and its Fourier transform. From Eq. (5.49), the scattering intensity is calculated as follows:

$$I(\vec{q}) = A(\vec{q})A(\vec{q})^* = \int_{V_2} \int_{V_1} \Delta\rho(\vec{r}_1)\Delta\rho(\vec{r}_2)e^{-i\vec{q}\cdot(\vec{r}_1 - \vec{r}_2)} \, d\vec{r}_1 \, d\vec{r}_2 \tag{5.56}$$

Replacing the variable $\vec{r} = \vec{r}_1 - \vec{r}_2$, Eq. (5.56) becomes

$$I(\vec{q}) = \int \left[\int_{V'} \Delta\rho(\vec{r}')\Delta\rho(\vec{r}' + \vec{r}) d\vec{r}' \right] e^{-i\vec{q}\cdot\vec{r}} \, d\vec{r} \tag{5.57}$$

or

$$I(\vec{q}) = V \int_{V'} \gamma(\vec{r}) \cdot e^{-i\vec{q}\cdot\vec{r}} \, d\vec{r} \tag{5.58}$$

where V is the irradiated sample volume and $\gamma(r)$ is the correlation function, defined as

$$\gamma(\vec{r}) = \frac{1}{V} \int_V \Delta\rho(\vec{r}')\Delta\rho(\vec{r}' + \vec{r}) d\vec{r}' \tag{5.59}$$

From Eqs. (5.57) and (5.58), it can be concluded that the scattering intensity defined in reciprocal space, $I(\vec{q})$, depends on the electron density of the scattering nanoparticle through its correlation function $\gamma(\vec{r})$. This function is obtained directly from the scattering intensity $I(\vec{q})$ by calculating the inverse Fourier transform:

$$\gamma(\vec{r}) = \frac{1}{(2\pi)^3 V} \int I(\vec{q})e^{i\vec{q}\cdot\vec{r}} \, d\vec{q} \tag{5.60}$$

2.1.6 X-Ray Scattering by Isotropic Systems

The X-ray scattering intensity produced by a isotropic system composed of N identical nanoparticles in dilute solution—for instance, macromolecules in solution—exhibits the following features:

1. For a statistically isotropic system, the random orientation of nanoparticles in solution requires that the scattering intensity due to a single nanoparticle $I_1(\vec{q})$ be accounted for all orientations, and, thus, the intensity becomes a function that depends solely on the magnitude of $\vec{q}$, that is, the intensity $I_1(q)$ is independent of the direction of $\vec{q}$.
2. Due to the absence of spatial correlation between the instantaneous positions of the N nanoparticles in a dilute solution, the total scattering intensity is simply N times the intensity produced by one nanoparticle averaged for all orientations, $I_1(q)$, that is, $I(q) = NI_1(q)$.

According to feature (1), the orientational average of the exponential term in Eqs. (5.55) and (5.58) is given by (Debye, 1915)

$$\langle e^{-i\vec{q}\cdot\vec{r}} \rangle = \frac{\sin qr}{qr} \tag{5.61}$$

Thus, Eq. (5.55), which corresponds to a cluster of atoms, can be rewritten as

$$I_1(q) = \sum_{i=1}^{n} \sum_{j=1}^{n} f_i f_j \frac{\sin q r_{ij}}{q r_{ij}} \qquad (5.62)$$

On the other hand, when applied to the description of an arbitrary isotropic structure, Eq. (5.58) becomes

$$I(q) = V \int 4\pi r^2 \gamma(r) \frac{\sin qr}{qr} dr \qquad (5.63)$$

where V is the volume of the irradiated sample and $\gamma(r)$ is the average of $\gamma(\vec{r})$ [Eq. (5.59)] for all $\vec{r}$ vector orientations:

$$\gamma(r) = \left\langle \frac{1}{V} \int_{V'} \Delta\rho(\vec{r}')\Delta\rho(\vec{r}' + \vec{r}) d\vec{r}' \right\rangle_\Omega \qquad (5.64)$$

As noted, both functions $\gamma(r)$ and $I(q)$ depend only on the magnitudes of the $\vec{r}$ and $\vec{q}$ variables, respectively. This result reflects the isotropic character of the analyzed structure and the intensity distribution in reciprocal space of the associated scattering intensity.

2.2 Nanoparticles Immersed in Homogeneous Matrices

2.2.1 Redefinition of the Correlation Function

The simplest model of nanostructured materials for the analysis of experimental SAXS results is the "two-electron density" model, characterized by the electron densities ρ_1 and ρ_0. This model is commonly applied to the study of many types of inorganic materials (e.g., nanoporous materials, metallic alloys containing nanoprecipitates, and colloidal solutions of nanoparticles) as well as organic materials such as nanostructured polymers and proteins in solution.

For isotropic materials modeled by two electron densities ρ_0 and ρ_1, the correlation function $\gamma(r)$ [Eq. (5.64)] is written as follows:

$$\gamma(r) = (\Delta\rho)^2 \gamma_0(r) \qquad (5.65)$$

where $\Delta\rho = (\rho_1 - \rho_0)$ and $\gamma_0(r)$ is the so-called "characteristic function." In the particular case of a dilute, monodisperse nanoparticle system, $\gamma_0(r)$ depends exclusively on the geometry (size and shape) of the nanoparticle.

For a single nanoparticle with volume V_1 immersed in a matrix with spatially constant density, it can be verified that (1) the characteristic function $\gamma_0(r)$ is a decreasing function with a maximum value of 1 for $r = 0$, (2) the integral of the function $\gamma_0(r)$ in three dimensions is equal to the particle volume V_1, and (3) $\gamma_0(r)$ vanishes for values of r equal to or greater than the maximum diameter of the nanoparticle, D_{max}, that is, $\gamma(r \geq D_{max}) = 0$.

In the specific case of a single nanoparticle with volume V_1 and maximum diameter D_{max}, explicit use of the characteristic function $\gamma_0(r)$ gives

$$I_1(q) = (\Delta\rho)^2 V_1 \int_0^{D_{max}} 4\pi r^2 \gamma_0(r) \frac{\sin qr}{qr} dr \qquad (5.66)$$

$$I_1(0) = (\Delta\rho)^2 V_1 \int_0^{D_{max}} 4\pi r^2 \gamma_0(r) dr = (\Delta\rho)^2 V_1^2 \qquad (5.67)$$

$$\gamma_0(r) = \frac{1}{8\pi^3 (\Delta\rho)^2 V_1} \int_0^\infty 4\pi q^2 I_1(q) \frac{\sin qr}{qr} dq \qquad (5.68)$$

$$\gamma_0(0) = \frac{1}{8\pi^3 (\Delta\rho)^2 V_1} \int_0^\infty 4\pi q^2 I_1(q) dq \qquad (5.69)$$

From Eq. (5.69), and bearing in mind that $\gamma_0(0) = 1$, the integral in reciprocal space associated with a nanoparticle, Q_1, is given by

$$Q_1 = \int_0^\infty 4\pi q^2 I_1(q) dq = 8\pi^3 (\Delta\rho)^2 V_1 \qquad (5.70)$$

2.2.2 Proteins in Solution

This subsection describes procedures for determining the shape and sizes of proteins in dilute solution. The relevant features of protein solutions to be considered are: (1) the set is monodisperse, that is, all proteins are identical; (2) there is no spatial correlation between their instantaneous positions; (3) the set is isotropic, that is, the proteins adopt all possible orientations; (4) the protein electron density is spatially constant; and (5) the proteins are immersed in a liquid medium (buffer) that also has a constant electron density. One of the procedures discussed in the subsequent text also allows for the study of proteins with a heterogeneous electron density.

In dilute solutions, the average electron density of the solution can be considered equal to that of the solvent ρ_0. The second electron density of the two-density model, ρ_1, is the average protein density. Thus, the relevant quantity related to the electron density of the proteins in solution is $\Delta\rho = (\rho_1 - \rho_0)$.

The diluted system condition ensures that each protein contributes independently (without interference effects) to the total scattering intensity. Thus, as discussed earlier, the total intensity produced by N proteins is simply N times the intensity produced by an individual protein averaged for all its orientations. Because all proteins are identical, only one protein needs to be considered, with all of its orientations.

From the characteristic function $\gamma_0(r)$ associated with a protein, the "distance distribution function" $p(r)$, defined as $p(r) = 4\pi r^2 \gamma_0(r)$, is derived. The orientational average of the scattering intensity by a protein $I_1(q)$ [Eq. (5.66)] can be rewritten as a function of $p(r)$:

$$I_1(q) = (\Delta\rho)^2 V_1 \int_0^{D_{max}} p(r) \frac{\sin qr}{qr} dr \qquad (5.71)$$

Consequently, the distance distribution function $p(r)$ is determined by

$$p(r) = \frac{1}{8\pi^3(\Delta\rho)^2 V_1} \int I_1(q) \frac{\sin qr}{qr} dq \qquad (5.72)$$

The function $p(r)$ is positive, equals 0 for $r = 0$, has one or more maxima, and decays to 0 for an r values larger than the maximum diameter of the protein, D_{max}. An initial analysis of protein geometry is generally performed using the function $p(r)$, determined by a inverse Fourier transform of the experimental intensity.

The general characteristics of the shape and size of proteins in solution can be deduced from a visual assessment of $p(r)$. Fig. 5.11 shows characteristic SAXS patterns and $p(r)$ functions for various geometric objects with the same value of D_{max}. Globular objects have a function $p(r)$ with an approximately Gaussian shape, with symmetry around the mean and a maximum at $r \approx D_{max}/2$. Elongated objects exhibit a $p(r)$ function with a maximum at smaller values of r, which approximately correspond to the cross-sectional radius R_c. Flattened objects show a much broader $p(r)$ maximum, shifted to distances smaller than $D_{max}/2$. A $p(r)$ function with a maximum shifted to distances larger than $D_{max}/2$ is indicative of a spherical shell shape. The $p(r)$ functions associated with objects composed of two separate subunits have two maxima:

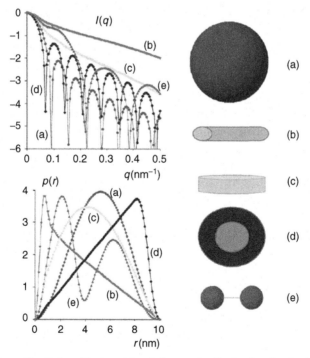

FIGURE 5.11 Scattering intensities (in logarithmic scale) and distance distribution function for different geometric objects. *Figure extracted from Svergun DI, Koch MHJ: Small-angle scattering studies of biological macromolecules in solution. Rep Prog Phys 66:1735–1782, 2003.*

one corresponds to intrasubunit distances and the other to intersubunit distances (Svergun and Koch, 2003).

2.3 Structural Parameters and Models

The SAXS technique allows researchers from various fields, from materials science to molecular biology, to determine structural parameters of nanoobjects in a homogeneous solution (liquid or solid). The following subsections describe the analytical procedures applied to SAXS results to determine the structural parameters of nanoobjects.

2.3.1 Determination of the Radius of Gyration of Nanoparticles in Dilute Solution

As previously mentioned, the total scattering intensity of a dilute, isotropic solution of N identical, randomly shaped nanoparticles (e.g., proteins in solution) is given by

$$I(q) = N \cdot I_1(q) \tag{5.73a}$$

For $q \to 0$, Guinier demonstrated that the SAXS intensity $I(q)$ can be approximated as a Gaussian function (Guinier and Fournet, 1955):

$$I(q) = NI_1(0)e^{-(R_g^2/3)q^2} \tag{5.73b}$$

where R_g is the radius of gyration with respect to the center of mass of the nanoparticles, defined by

$$R_g = \left(\frac{1}{n_e} \int_{V_1} r^2 \rho(\vec{r}) d\vec{r} \right)^{1/2} \tag{5.74}$$

where n_e is the total number of electrons in the particle. For a two-electron density model of N homogeneous nanoparticles per unit volume with electron density ρ_1 immersed in a matrix with density ρ_0, Eqs. (5.73b) and (5.74) can be written as

$$I(q) = N(\Delta\rho)^2 V_1^2 \, e^{-(R_g^2/3)q^2} \quad \text{and} \quad R_g = \left(\frac{1}{V_1} \int_{V_1} r^2 \, d\vec{r} \right)^{1/2} \tag{5.75}$$

respectively, wherein $\Delta\rho$ is the difference $\rho_1 - \rho_0$ (e.g., the difference between the average density of the proteins in solution and the density of the buffer) and V_1 is the volume of the nanoparticle.

For a homogeneous spherical nanoparticle with radius R, $R_g = \sqrt{3/5}R$. According to Eqs. (5.73a) and (5.73b), a Guinier plot, $\ln I(q) \times q^2$, should show a straight line with a negative slope at small values of q. The slope of this straight line is given by $\alpha = -R_g^2/3$, and R_g can in turn be calculated using $R_g = \sqrt{-3\alpha}$. As mentioned Guinier law [Eqs. (5.73a), (5.73b), and (5.75)] is valid only for values of q close to 0. For example, for approximately isodiametric nanoparticles, the upper limit of q for which the Guinier law remains valid is approximately equal to $1.3/R_g$.

2.3.2 Determination of the Surface/Volume Ratio of Nanoparticles

For sets of monodispersed nanoparticles that exhibit well-defined interfaces (e.g., proteins in solution) and for isotropic nonparticulate (bicontinuous) systems that can be modeled by two electron densities, the asymptotic behavior of $I(q)$ at high values of q is described by Porod's law (Glatter and Kratky, 1982). For the particular case of a isotropic, monodisperse and dilute set of N nanoparticles, Porod's law is written as

$$I(q) = \frac{2\pi(\Delta\rho)^2 N \cdot S_1}{q^4} \quad (q \to \infty) \tag{5.76}$$

where S_1 is the area of the interface between one nanoparticle and the matrix. In more general cases the asymptotic value $I(q)q^4$ for $q \to \infty$ is proportional to the total area of the interface, S, between the two phases of the system.

Furthermore, from Eqs. (5.70) and (5.76), and given that $Q = N \cdot Q_1$, the ratio between the interface area S_1 and volume V_1 of the nanoparticles is given by

$$\frac{S_1}{V_1} = \frac{4\pi^2}{Q}\left[\lim_{q\to\infty} q^4 I(q)\right] \tag{5.77}$$

Electron density fluctuations in the nanoparticles and/or in the matrix produce a constant contribution to the scattering intensity. To suppress the effect of electron density fluctuations in the SAXS intensity curves, plots of $q^4 I(q)$ versus q^4 usually exhibit a linear dependence for large values of q, $I(q)q^4 = A + Bq^4$. From these plots, the linear (A) and angular (B) coefficients are determined. Thus, the difference function $[I(q) - B]$ at the limit $q \to \infty$ satisfies Porod's law: $[I(q) - B] = A/q^4$. Then, the coefficient A is used, for example, to calculate the interface area S using Eq. (5.76) or the nanoparticle ratio S_1/V_1 using Eq. (5.77).

2.3.3 Determination of the Volume of Nanoparticles in Dilute Solution

Considering a two-electron density model for a dilute, monodisperse, randomly oriented (proteins or others) nanoparticle system, the volume V_1 of the nanoparticles can be determined by measuring the small-angle scattering intensity on a relative scale. For this purpose, it is necessary to determine the scattering intensity at $q = 0$, $I(0)$ (by adequate extrapolation using Guinier law), and the integral of $I(q)$ in reciprocal space, Q. From Eqs. (5.67), (5.70), and (5.73a), the nanoparticle volume is given by

$$V_1 = 8\pi^3 \frac{I(0)}{Q} \tag{5.78}$$

In turn, the value of Q is determined by calculating the integral of $I(q)$ over the entire reciprocal space volume ($Q = \int 4\pi q^2 I(q) dq$), for which the small-angle scattering intensity does not vanish. In typical SAXS experiments, the maximum accessible q value, q_{max}, ranges between 0.3 and 0.8 Å^{-1}. To complete the integration of $I(q)$ for values of $q > q_{max}$, the $I(q)$ curve is usually extrapolated using Porod's law, after subtracting the constant contribution due to density fluctuations, as mentioned in the precedent subsection.

2.4 Software for the Analysis of SAXS Curves of Proteins in Solution

2.4.1 Distance and Size Distribution Functions

The inverse Fourier transform of the SAXS intensity, $I(q)$, is calculated as specified in Eq. (5.72) to generate the distance distribution function, $p(r)$. This mathematical operation is not simple due to undesirable effects produced by statistical errors in the measurement of $I(q)$, which are important for large q values, and due to the cutting of function $I(q)$ for $q = q_{max}$. Therefore, $p(r)$ is generally determined using programs that minimize the aforementioned effects.

One of the most commonly used programs for determining $p(r)$ is named GNOM (Svergun, 1992). The inputs are the SAXS intensity curve—after subtracting the intensity of parasitic scattering—and the statistical errors of measurement (the inclusion of statistical errors is not essential). Through a method of indirect transformation, GNOM usually outputs a stable $p(r)$ curve, without spurious oscillations. The GNOM program also estimates the quality of the outcome.

In addition to generating $p(r)$ functions corresponding to identical nanoparticles (proteins or other nanoparticles) in dilute solution, GNOM can also be used to determine the size distribution of polydisperse sets of nanoparticles of same shape in dilute solution. The program determines this distribution for nanoobject sets with simple geometric shapes, such as spheres, cylinders, and prisms.

Finally, an important feature of GNOM is its ability to account for the effects of SAXS curve distortions (smearing) arising from: (1) the size of the cross-section of the incident X-ray beam, (2) the pixel size of the X-ray detector, and (3) the spectral width (in energy or wavelength) of the incident beam. The latter effect is negligible in typical SAXS measurements, but is usually important in small-angle neutron scattering (SANS) experiments. In typical experiments using synchrotron X-ray sources, the cross-section of the incident beam and the detector pixel size are both very small, thus making unnecessary, in many cases, the use of mathematical desmearing procedures.

2.4.2 SAXS Curves Determined From High-Resolution Structures

Comparisons of experimental SAXS curves and curves simulated from atomic resolution structures—obtained by single-crystal XRD or nuclear magnetic resonance (NMR)—have frequently been used to validate theoretical models, to verify structural differences between proteins and nucleic acids in a crystalline state and in solution, and to predict the quaternary structures of macromolecules. High-resolution structures of many proteins are deposited and are accessible in the Protein Data Bank (PDB, www.rcsb.org)

Proteins in solution can be schematically represented as shown in Fig. 5.12. A protein with electron density $\rho_a(\vec{r})$ is surrounded by a solvent with electron density ρ_0. Hydration of the protein is represented by a layer of thickness Δ and electronic density ρ_b covering its entire surface. As stated previously, the SAXS curve of a set of N proteins per unit volume

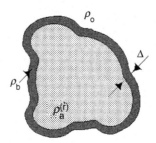

FIGURE 5.12 Schematic low-resolution representation of a protein in solution.

immersed in a sufficiently diluted solution is proportional to the average value of a single protein's scattering intensity for all of its orientations. Thus, the scattering intensity per unit volume can be written as

$$I(q) = N\langle| A_a(q) - \rho_s A_s + \delta\rho A_b(q)|^2\rangle_\Omega \qquad (5.79)$$

where $A_a(q)$, $\rho_s A_s$, and $\delta\rho A_b(q)$ are the amplitudes of the beam scattered by each protein in vacuo, by the excluded solvent volume, and by the hydration layer, respectively, with $\delta\rho = \rho_b - \rho_s$.

For proteins whose high-resolution structures have been determined by single-crystal XRD or NMR, the SAXS intensity curves produced by a set of identical proteins in dilute solution can be determined using the multipole expansion method in the CRYSOL program (Svergun et al., 1995). CRYSOL is usually applied to determine the scattering intensity $I(q)$ from the high-resolution protein structures extracted from the PDB. Relevant parameters are determined by fitting the intensity $I(q)$ to the experimental curve $I_e(q)$. These parameters are the electron density difference between the hydration layer and the solvent, $\delta\rho$, and the radius of the dummy atom r_0 used by the program to define the excluded solvent volume. The aforementioned fitting is performed by minimizing the discrepancy:

$$\chi^2(r_0, \delta\rho) = \frac{1}{N_p}\sum_{i=1}^{N_p}\left[\frac{I_e(q_i) - cI(q_i, r_0, \delta\rho)}{\sigma(q_i)}\right]^2 \qquad (5.80)$$

where N_p is the number of points in the experimental curve $I_e(q_i)$, $\sigma(q_i)$ represents the statistical errors of the intensity measurements, and c is a scaling factor.

2.4.3 Ab Initio Determination of Protein Shape From Experimental SAXS Curves

The challenge of applying SAXS to structural studies of proteins in solution lies in determining their shape and size using a three-dimensional model based on a one-dimensional function—the scattering curve $I(q)$. However, except for trivial cases of spherical particles, it cannot be stated with certainty that there is a single solution to the problem, as the $I(q)$ functions determined for different models can, in principle, show similar discrepancies in relation to the experimental curves (Volkova and Svergun, 2003).

In the past, the protein shape and size were usually determined by calculating the intensity of SAXS curves produced by three-dimensional models of objects with simple geometric shapes such as spheres, cylinders, ellipsoids, and prisms "(averaged for all orientations) and comparing these curves with those obtained experimentally.

The first ab initio method for determining protein shape was proposed by Stuhrmann (1970a, 1970b). In this method, the protein is represented by an angular envelope function that describes particles in spherical coordinates. However, the use of the angular envelope function is limited to relatively simple protein shapes.

Currently, the three-dimensional dummy atom model (DAM) is commonly applied. This model consists of a cluster of small spheres (dummy atoms) that is applied to the study of homogeneous proteins and also heterogeneous proteins comprising several phases. This method was first implemented in the DALAI-GA program (Chacón et al., 1998, 2000). The initial model consists of, for example, a sphere of radius R equal to half the maximum diameter of the protein, D_{max} (usually determined using GNOM). The initial sphere contains several phases with different electron densities and is filled with a dense packing of smaller dummy atoms with radius r_0. Thus, the total number of dummy atoms is approximately given by $N = (R/r_0)^3$.

Each dummy atom in DAM is identified with an index X_j indicating the phase to which it belongs. The index X_j adopts values ranging from 0 (for the solvent) to k (for the maximum protein electron density). The coordinates of the positions for each dummy atom complete the low-resolution characterization of the studied protein.

Each dummy atom of the kth phase exhibits a density contrast $\Delta\rho_k$, and the total scattering intensity for the DAM is given by

$$I(q) = \left\langle \left[\sum_{k=1}^{K} \Delta\rho_k A_k(\vec{q}) \right]^2 \right\rangle_{\Omega} \tag{5.81}$$

where $A_k(q)$ is the scattering amplitude produced by the volume occupied by the kth protein phase and $< >_{\Omega}$ indicates orientational averaging. The amplitude function is represented by spherical harmonics $Y_{lm}(\Omega)$:

$$A_k(q) = \sum_{l=0}^{\infty} \sum_{m=-1}^{1} A_{lm}^{(k)}(q) Y_{lm}(\Omega) \tag{5.82}$$

where $A_{lm}^{(k)}(q)$ are the partial amplitudes. The partial amplitudes for the volume occupied by the kth phase in the DAM are given by the sum of all N_k dummy atoms in that phase:

$$A_{lm}^{(k)}(q) = i^l \sqrt{\frac{2}{\pi}} f(q) \sum_{j=1}^{N_k} j_1(qr_j) Y_{lm}^*(\omega_j) \tag{5.83}$$

where r_j and ω_j are polar coordinates, $j_1(qr_j)$ is the first-order Bessel function, and $f(q)$ is the scattering factor of the dummy atom. Thus, the total intensity is given by

$$I(q) = 2\pi^2 \sum_{l=0}^{\infty} \sum_{m=-1}^{1} \left[\sum_{k=1}^{K} [\Delta\rho_k A_{lm}^{(k)}(q)]^2 + 2\sum_{n>k} \Delta\rho_k A_{lm}^{(k)}(q) \Delta\rho_n [A_{lm}^{(n)}(q)]^* \right] \tag{5.84}$$

The DAMMIN program (Svergun, 1999) assumes a single electron density throughout the entire protein volume, which, in many cases, is a good approximation. The task of obtaining a low-resolution model from a SAXS curve consists of finding a configuration of dummy atoms X, from the initial DAM, that minimizes the function

$$f(X) = \chi^2 + \alpha P(X) \tag{5.85}$$

where χ^2 is the discrepancy between the simulated curve $I(q)$ and the experimental curve $I_{exp}(q)$, $P(X)$ is a "penalty" function that aims to reject solutions with unacceptable configurations—such as those that exhibit a loss of connectivity between dummy atoms—and α is a positive parameter that defines the relative effect of function $P(X)$.

The discrepancy parameter χ^2 between the experimental scattering intensity curve, $I_{exp}(q)$, and the scaled scattering intensity curve calculated for the DAM, $I(q)$, is given by

$$\chi^2 = \frac{1}{n} \sum_{j=1}^{n} \left[\frac{I_{exp}(q_j) - I(q_j)}{\sigma(q_j)} \right]^2 \tag{5.86}$$

where n is the number of points in the experimental function $I_{exp}(q)$ and $\sigma(q_j)$ denotes the statistical errors in $I_{exp}(q_j)$.

The three-dimensional shape of the protein is obtained as the end result of the minimization process of $f(X)$. During minimization, difficulties in achieving the absolute minimum of function $f(X)$ arise due to the presence of relative minima. To address this issue, DAMMIN employs a method known as simulated thermal annealing (Kirkpatrick et al., 1983).

To find a low-resolution structural model of a protein in solution, the GASBOR program (Svergun et al., 2001) may be used in addition to DAMMIN. Proteins typically consist of coiled polypeptide chains composed of amino acid residues with a separation of approximately 0.38 nm between adjacent alpha carbons. The GASBOR program models the structure through dummy polypeptide chain residues called dummy residues (DRs). The three-dimensional shape of the protein is determined by finding the arrangement of DRs compatible with the primary chain whose simulated SAXS curve fits the experimental scattering pattern.

2.4.4 Determination of the Molecular Mass of Proteins From Experimental SAXS Curves on a Relative Scale

The molecular mass of proteins, M_m, can be determined from the extrapolated scattering intensity I(0), determined on absolute scale. However, the determination of $I(q)$ on an absolute scale requires a knowledge of the incident X-ray beam intensity, whose measurement is inaccurate and generally performed using calibrated standard samples. Another procedure compares the $I(0)$ values of a diluted solution containing the protein of interest and of a solution with a protein whose M_m is known. This method requires knowledge of the concentrations of the two solutions, which cannot always be accurately determined.

An alternative method that requires only intensity measurements on a relative scale up to the maximum value of q (q_{max}) within the range used in typical SAXS experiments was proposed by Fischer et al. (2010). The method determined a number of SAXS intensity curves by applying CRYSOL, corresponding to 1148 high-resolution protein structures with known real volumes V_{real} (extracted from the PDB). From these curves, apparent volumes are calculated using the equation $V_{ap} = 8\pi^3 I(0)/Q_P$, where Q_P refers to partial integrals determined up to different q_{max} values. With the accurately determined real volumes of these proteins, calculated from known high-resolution structures, several (approximately linear) V_{real} versus V_{ap} functions can be determined for different values of q_{max}.

The steps of the proposed procedure are as follows: (1) determine the relative intensity $I(0)$ by adequate extrapolation of $I(q)$ using Guinier law; (2) calculate the partial integral Q_{ap}, also on a relative scale, until reaching the value of q_{max} used in the experiment; (3) calculate the apparent volume of the protein, $V_{ap} = I(0)/Q_{ap}$; (4) determine the value of V_{real}, the molecular volume of the protein studied, using the known function (V_{real} vs. V_{ap}) for the q_{max} value used in the experiment; and (5) determine the molecular mass $M_m = V_{real} \cdot \rho_P$, using a protein density value ρ_P of $1.37\ \text{g/cm}^3$ (Squire and Himmel, 1979).

2.5 Example of Application: Study by SAXS of *Leptospira* Ferrodoxin-NADP(H) Reductase in Solution

The protein named Leptospira Ferrodoxin Reductase (*Lep*FNR) consists of two domains. The C-terminal domain is highly conserved among FNRs, consisting of an α/β sandwich with five β-sheets surrounded by six α-helices, while the N-terminal domain has six antiparallel β-sheets to which the FAD cofactor binds. FNRs are known as monomeric proteins present in plastids, bacteria, mitochondria, and apicoplasts of intracellular parasites. However, it was recently suggested that two *Arabidopsis thaliana* FNR isoforms are able to dimerize (Lintala et al., 2007).

The *Lep*FNR protein complexed with its FAD cofactor was studied in dilute solution using SAXS to determine its shape and oligomerization state (Nascimento et al., 2007). Experimental $I(q)$ curves of *Lep*FNR at concentrations of 3 and 10 mg/mL were measured using the SAXS line of the Brazilian Synchrotron Light Laboratory (LNLS) (Kellermann et al., 1997). The X-ray wavelength used was 0.148 nm. A position-sensitive one-dimensional X-ray detector was used to record the scattering intensity as a function of the modulus of the scattering vector q. Parasitic X-ray scattering produced by the sample cell windows and by the incident beam-defining slits was suppressed by subtracting a SAXS intensity curve produced by the buffer from the total intensity curve. The sample–detector distance was 1155 mm. Thus, the X-ray scattering intensity was recorded in the range of $0.1\ \text{nm}^{-1} < q < 3.0\ \text{nm}^{-1}$.

The SAXS curves produced by *Lep*FNR at concentrations of 3 and 10 mg/mL exhibit the same scattering profiles, indicating that the solution can be considered diluted up to a concentration of at least 10 mg/mL. Thus, the $I(q)$ curve corresponding to the protein at $c = 10$ mg/mL was used for further data analysis.

Various analytical procedures were applied to the experimental SAXS intensity curve (Fig. 5.13), corresponding to the *Lep*FNR protein complexed with its FAD cofactor in dilute

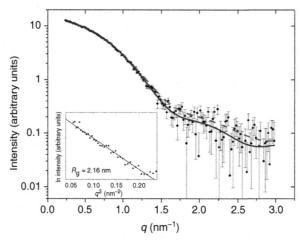

FIGURE 5.13 Experimental SAXS curve (*circles with error bars*), DAM-simulated scattering intensity (*dashed line*), simulated scattering intensity (*solid line*) derived from the high-resolution structure of LepFNR (PDB id:2RC5). (Inset) Guinier plot from which R_g was calculated. *DAM*, Dummy atom model; *LepFNR, Leptospira* ferredoxin-NADP(H) reductase; *PDB*, Protein Data Bank; *SAXS*, small-angle X-ray scattering. *Figure from Nascimento AS, Catalano-Dupuy DL, Bernardes A, Oliveira Neto M, Santos MAM, Ceccarelli EA, Polikarpov I: Crystal structures of Leptospira interrogans FAD-containing ferredoxin-NADP+ reductase and its complex with NADP+BMC, Struct Biol 7:69–81, 2007.*

solution, whose high-resolution structure is known. The following parameters and structural characteristics of this protein in solution were determined:

1. From the straight line slope in the Guinier plot shown in Fig. 5.13, the radius of gyration was calculated as $R_g = 2.16$ nm.
2. Using GNOM, the distance distribution function $p(r)$ shown in Fig. 5.14 was determined, which gave a maximum diameter $D_{max} = 6.5$ nm and a radius of gyration $R_g = 2.13$ nm.

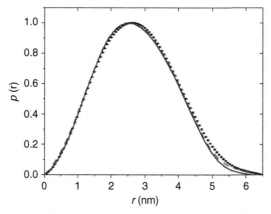

FIGURE 5.14 Distance distribution functions $p(r)$ for *LepFNR*, calculated using (1) GNOM program with the experimental $I(q)$ curve as input (*circles*), (2) DAM (*dashed line*), and (3) high-resolution *LepFNR* monomer (PDB id: 2RC5) (*solid line*). *DAM*, Dummy atom model; *LepFNR, Leptospira* ferredoxin-NADP(H) reductase; *PDB*, Protein Data Bank. *Figure from Nascimento AS, Catalano-Dupuy DL, Bernardes A, Oliveira Neto M, Santos MAM, Ceccarelli EA, Polikarpov I: Crystal structures of Leptospira interrogans FAD-containing ferredoxin-NADP+ reductase and its complex with NADP+BMC, Struct Biol 7:69–81, 2007.*

3. Using the procedure of Fischer et al., the molecular mass was determined to be $M_m = 35.4$ kDa.

4. GASBOR was utilized to determine the DAM or low-resolution structure, which is shown in Fig. 5.15 superimposed on the high-resolution structure extracted from the PDB. The DAM values of R_g and D_{max} are 1.97 and 6.00 nm, respectively.

5. Using CRYSOL and the high-resolution structure extracted from the PDB, the SAXS intensity was calculated, as shown in Fig. 5.13 along with the experimental curve.

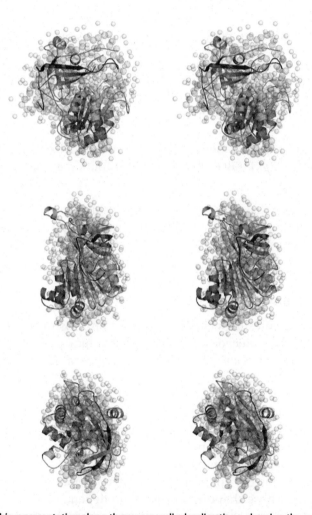

FIGURE 5.15 **Stereographic representation along three perpendicular directions, showing the superposition of the** *Lep***FNR protein monomer structure (extracted from PDB) with the ab initio DAM obtained from the SAXS curve using the GASBOR program.** *DAM,* Dummy atom model; *Lep*FNR, *Leptospira* ferredoxin-NADP(H) reductase; *PDB,* Protein Data Bank; *SAXS,* small-angle X-ray scattering. *Figure from Nascimento AS, Catalano-Dupuy DL, Bernardes A, Oliveira Neto M, Santos MAM, Ceccarelli EA, Polikarpov I: Crystal structures of* Leptospira interrogans *FAD-containing ferredoxin-NADP+ reductase and its complex with NADP+BMC,* Struct Biol 7:69–81, 2007.

Table 5.6 Structural Parameters of the *LepFNR* Protein Obtained by the Analysis of SAXS Curves and from the High-Resolution Structure Extracted from the Brookhaven Protein Data Bank (PDB)

Parameter	Guinier	GNOM	DAM	Fischer et al.	High-Resolution Model
D_{max} (nm)		6.50	6.00		6.14
R_g (nm)	2.16	2.13	1.97		1.93
M_m (kDa)				35.4	34.4

DAM, dummy atom model; *LepFNR*, *Leptospira* ferredoxin-NADP(H) reductase; *SAXS*, small-angle X-ray scattering.

The aforementioned results allow the following conclusions:

1. There is good agreement between the values of R_g, M_m, and D_{max} of *LepFNR* determined by various methods (Table 5.6).
2. There is a clear similarity between the $I(q)$ curves determined experimentally and those obtained using CRYSOL from the high-resolution structure of the protein monomer.
3. There is also a clear similarity between the ab initio model generated from the experimental $I(q)$ using GASBOR and the high-resolution monomer structure (Fig. 5.15).

These results indicate that the *LepFNR* protein in solution is certainly in a monomeric state and that, in solution, it retains essentially the same shape as in the crystalline state.

3 GISAXS and ASAXS

3.1 Grazing-Incidence X-Ray Scattering

The study of nanostructures deposited in or grown on substrates has increased significantly during the past two decades, as these systems may present unique features. Interest in these systems is largely due to their technological importance in the catalysis (Kasai and Escaño, 2016; Lambert and Pacchioni, 1997), protective coating (Voevodin et al., 2004; Zhang, 2010), lubricant (Mortier et al., 2010), semiconductor (Fendler, 1998; Gaponenko, 1998; Peng and Mingos, 2005), and solar cell (Borchert, 2014) industries, among others. Therefore, research on the physical properties of surfaces and interfaces, as well as their correlations with morphology and structure obtained after a given preparation process, is fundamentally important.

A particular example is the epitaxial growth of small nanoscale crystals over monocrystalline substrates, whose shape and orientation depend on, among other characteristics, the structure and crystallographic facets over which they are deposited (Liakakos et al., 2015; Peng et al., 1997). Another example consists of the formation of ordered structures within single crystals, which originate after the diffusion and aggregation of metal atoms contained in the overlying thin films, giving rise to ordered nanostructures with uniform size and shape (Kellermann et al., 2012). Such materials have properties with potential application in the construction of miniaturized optical or electronic devices, with low energy consumption and high switching speed, among others (Klimov, 2005; Nazarov et al., 2014). However, understanding the processes that lead to nanoparticle-containing materials with certain characteristics—such as size, shape, order, and size dispersion—that confer useful properties requires characterization techniques that allow one to monitor the formation and growth of these nanostructures.

To study these structures by SAXS, the conventional transmission geometry—in which X-rays must traverse the sample—is not appropriate because the substrates on which these films are deposited often strongly absorb X-rays, preventing them from being transmitted through the sample. Furthermore, the SAXS intensity in this geometry is generally low, as the irradiated volume of thin films is small. Furthermore, the substrate may also contribute to the total X-ray scattering. In most cases, because the substrate thickness is much larger than the film thickness, scattering from the substrate may be significantly larger than the scattering from the nanostructures in the film. For these reasons, the most suitable experimental setup for studying thin films is one in which the primary beam hits the sample surface at a grazing angle. Thus, the structures close to the surface will scatter X-rays, which will only be partially absorbed by the film layer above them (if said layer is present). The technique that makes use of this geometry is known as grazing-incidence small-angle X-ray scattering (GISAXS) (Ezquerra et al., 2009; Pietsch et al., 2004).

Unlike conventional SAXS geometry, which does not distinguish between intensities originating from the surface and those arising from deeper layers of the material, the GISAXS technique allows one to obtain mainly the signal from structures deposited on the surface or buried within a thin layer close to the surface. The thickness of the studied layer can be adjusted by choosing a suitable incidence angle and X-ray photon energy. The next section will describe how these values vary for different materials, based on the dependence of the refractive index on photon energy and sample composition.

The analysis of GISAXS intensities is performed in a manner similar to that described for the general SAXS theory in the previous section; however, in the case of GISAXS, the reflection and refraction of X-rays at the interfaces must be taken into account when calculating the scattered intensity. The following section briefly presents this technique, highlighting the main aspects that distinguish it from the aforementioned conventional transmission technique.

3.1.1 Refractive Index, Penetration Depth, and Fresnel Reflection and Transmission

This section will review some basic concepts involving the X-ray reflection and refraction processes and their influence on the selection of an adequate experimental setup for the study of different types of thin films. The following sections will also describe how X-ray reflection and refraction at various sample interfaces have a decisive influence on the scattering intensity pattern produced by the sample.

3.1.1.1 REFRACTIVE INDEX

The refractive index or index of refraction is a complex number given by (Tolan, 1999)

$$n = 1 - \delta(\vec{r}) + i\beta(\vec{r}) \tag{5.87}$$

where

$$\delta(\vec{r}) = \frac{\lambda^2}{2\pi} r_e \rho(\vec{r}) \sum_{j=1}^{N} \frac{f_{0j} + f_j'(\lambda)}{Z} \tag{5.88}$$

and

$$\beta(\vec{r}) = \frac{\lambda^2}{2\pi} r_e \rho(\vec{r}) \sum_{j=1}^{N} \frac{f_j''(\lambda)}{Z} = \frac{\lambda}{4\pi} \mu(\vec{r}) \tag{5.89}$$

In Eqs. (5.88) and (5.89), λ is the X-ray wavelength, $r_e = 2.814 \times 10^{-5}$ Å is the classical electron radius, $\rho(\vec{r})$ is the number of electrons per unit volume at position $\vec{r}$ of the material, μ is the linear absorption coefficient, f_{0j} is the atomic scattering factor for element j for photon energies far from those corresponding to absorption edges, and f_j' and f_j'' are the real and imaginary parts, respectively, of the "anomalous" (or resonant) components of the atomic scattering factor. These correction terms are particularly important when the X-ray energy is close to the absorption edges of the elements present in the sample. For X-rays, values of δ and β are typically on the order of 10^{-5} and 10^{-7}, respectively. $Z = \sum_i Z_i$ is the total number of electrons, wherein Z_i is the number of electrons in element i.

By impinging on the sample surface at an angle α_i greater than or equal to the critical angle of total external reflection α_c, the monochromatic X-ray beam splits into a reflected beam and a refracted beam (Tolan, 1999). This situation is shown schematically in Fig. 5.16. The relationship between α_i and the angle of refraction α_t is given by Snell's law: $n_i \cos \alpha_i = n_t \cos \alpha_t$ (Als-Nielsen and McMorrow, 2011).

For X-ray photons, the real part of the refractive index of materials is smaller than the refractive index of air ($n_{air} \cong 1$) or vacuum ($n_{vacuum} = 1$). Therefore, by impinging on a flat surface with $\alpha_i < \alpha_c$, the X-ray beam coming from air (or vacuum) undergoes total external reflection, that is, the beam is almost entirely reflected and there is no significant fraction of photons transmitted through the surface. The critical angle of total external reflection, α_c, below which no refraction occurs and virtually the entire beam is reflected, can be calculated from Snell's law, taking $\alpha_t = 0$.

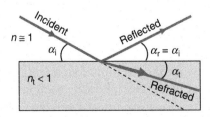

FIGURE 5.16 Schematic representing the incident, reflected, and refracted X-ray beams.

3.1.1.2 PENETRATION DEPTH

In GISAXS studies of buried nanoparticles, the refracted beam acts as a primary beam for the scattering process. The interaction between the primary beam and the nanostructures buried below the surface gives rise to a scattering signal from a depth that depends on the incidence angle and photon energy.

The parameter commonly used to quantify the penetration of X-rays in the sample is the penetration depth Λ, given by (Tolan, 1999)

$$\Lambda = \frac{\lambda}{\sqrt{2\pi}}\left\{\sqrt{\left(\alpha_i^2 - \alpha_c^2\right)^2 + 4\beta^2} - \left(\alpha_i^2 - \alpha_c^2\right)\right\}^{-1/2} \tag{5.90}$$

The magnitude of Λ represents the depth, in a direction perpendicular to the surface, at which the refracted beam intensity, I_t, decreases by a factor equal to $1/e$ of its value just above the surface. Fig. 5.17A shows how the $I_t(d)/I_t(0)$ ratio varies as a function of depth d for different values of Λ. The dependency of Λ in relation to α_i for various materials and for different X-ray photon energies is shown in Fig. 5.17B and C, respectively.

3.1.1.3 FRESNEL EQUATIONS FOR X-RAY REFLECTION AND TRANSMISSION

The I_t/I_i and I_r/I_i ratios, where I_t and I_r are the intensities of the refracted and reflected beams, respectively, and I_i is the intensity of the incident beam, can be determined using the Fresnel equations (Tolan, 1999). For the electric field vector perpendicular to the plane of incidence[a] (*s* polarization), the reflection and transmission coefficients for the X-ray intensities are, respectively, given by

$$R_S = \left|\frac{k_{iz} - k_{tz}}{k_{iz} + k_{tz}}\right|^2 \tag{5.91}$$

and

$$T_S = \left|\frac{2k_{iz}}{k_{iz} + k_{tz}}\right|^2 \tag{5.92}$$

where k_{iz} and k_{tz} are the components of the incident and refracted wave vectors, respectively, in the direction normal to the surface. The values of k_{iz} and k_{tz} are given by

$$k_{iz} = -k\sin\alpha_i \tag{5.93}$$

and

$$k_{tz} = -nk\sin\alpha_t = -k\sqrt{n^2 - \cos^2\alpha_i} \tag{5.94}$$

where k is the magnitude of the incident beam wave vector, given by $k = 2\pi/\lambda$. It is important to note that k_{tz} is a complex number.

For the electric field component parallel to the plane of incidence (*p* polarization), the reflection and transmission coefficients for the X-ray intensities are written as follows:

---

[a]The plane of incidence is the plane containing the incident beam and a straight line normal to the surface.

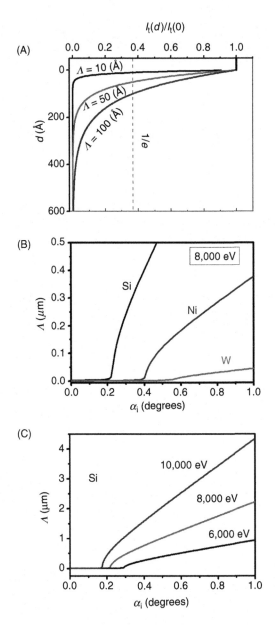

FIGURE 5.17 (A) $I_t(d)/I_t(0)$ ratio as a function of depth d for selected values of penetration depth Λ. $I_t(d)$ represents the intensity of the refracted beam at depth d, measured with respect to the air/sample interface, and $I_t(0)$ is the intensity of the refracted beam at the interface. (B) Λ versus α_i functions calculated for different materials using 8-keV photons. (C) Λ versus α_i functions for Si, calculated for different photon energy values.

$$R_p = \left| \frac{n^2 k_{iz} - k_{tz}}{n^2 k_{iz} + k_{tz}} \right|^2 \qquad (5.95)$$

and

$$T_p = \left| \frac{2 k_{iz}}{n^2 k_{iz} + k_{tz}} \right|^2 \qquad (5.96)$$

Fig. 5.18 shows a plot of the functions in Eqs. (5.91) and (5.92) for SiO_2 and silver at a photon energy of 8 keV. The values of α_c are indicated by vertical arrows. Note that for $\alpha_i < \alpha_c$, the X-ray transmission through the air/sample interface drops rapidly. Therefore, this condition is unsuitable for studying nanostructures buried within the substrate. For buried particles, it is necessary to use $\alpha_i > \alpha_c$. Once the depth of the nanoparticles is known, the incidence angle that maximizes the GISAXS intensity can be calculated from Eq. (5.90). It can be shown that $\alpha_c \cong \lambda \sqrt{r_e \rho / \pi}$, that is, α_c is proportional to the square root

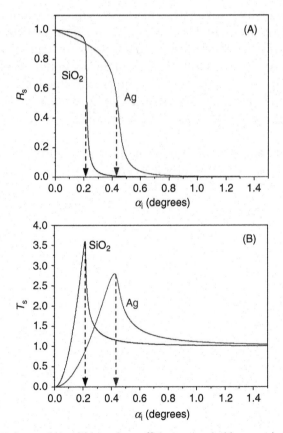

FIGURE 5.18 Reflection (A) and transmission (B) intensity coefficients versus incidence angle for SiO_2 and silver (Ag), calculated using Eqs. (5.91) and (5.92) for photons with an energy of 8 keV. The vertical arrows indicate α_c values.

of the electron density (Tolan, 1999) and therefore depends on the sample composition and density.

For nanoparticles deposited on the surface, it is more convenient to use $\alpha_i = \alpha_c$. In this case, the refracted beam propagates parallel to the surface ($\alpha_t = 0$), and the electromagnetic waves associated with the reflected and refracted beams interfere constructively, increasing the scattered intensity. The penetration of the resulting interference between the reflected and refracted beams, also known as an evanescent wave, is limited to a few nanometers below the surface (~ 5 nm in most materials)—eliminating (or strongly reducing) the contribution from nanostructures that may be buried within the substrate.

Finally, it is worth noting that the Fresnel reflection and transmission functions shown earlier do not consider the effects of air/sample interface roughness, which may be generally present. In such cases, it is necessary to use correction factors that depend on the model used to describe this roughness. A more detailed discussion can be found in several books on this subject (Tolan, 1999).

3.1.2 Scattering Vector

As noted earlier in this chapter, an analysis of the SAXS intensity requires knowledge of the scattered intensity as a function of the scattering vector $\vec{q}$. In a GISAXS experiment, the intensity of scattered X-rays is generally measured using some type of position-sensitive detector, which allows one to obtain the intensity of the scattered beam as a function of the position of incidence in the detector. It will be shown next that if the positions at which the incident and scattered beams intersect the detection plane and the sample-to-detector distance are known, the scattering vector $\vec{q}$ corresponding to that position can be easily determined.

Fig. 5.19 shows a schematic representation of the parameters used for determining $\vec{q}$. In this representation, $\vec{k}_i$ and $\vec{k}_f$ are the wave vectors of the incident and scattered beams, respectively, given by (Tolan, 1999)

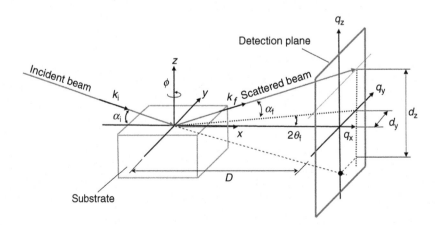

FIGURE 5.19 Parameters used for determining the scattering vector $\vec{q}$ in GISAXS experiment using a 2D X-ray detector.

$$\vec{k}_i = k(\cos\alpha_i, 0, -\sin\alpha_i)$$

(5.97)

$$\vec{k}_f = k(\cos 2\theta_f \cos\alpha_f, \ \sin 2\theta_f \cos\alpha_f, \ \sin\alpha_f)$$

(5.98)

where α_i and α_f are the angles between the incident and scattered beams and the sample surface, respectively, $2\theta_f$ is the angle between the projection of the scattered beam over the sample surface and the *x*-axis, and $k = 2\pi/\lambda$ is the magnitude of the wave vector. Consequently, the transferred momentum vector (or scattering vector) will be

$$\vec{q} = \vec{k}_f - \vec{k}_i = \frac{2\pi}{\lambda} \begin{pmatrix} \cos 2\theta_f \cos\alpha_f - \cos\alpha_i \\ \sin 2\theta_f \cos\alpha_f \\ \sin\alpha_f + \sin\alpha_i \end{pmatrix}$$

(5.99)

Knowing the value of α_i and the position of incidence of the direct beam in the detector, the angles α_f and $2\theta_f$ can be calculated using the following equations:

$$\alpha_f = \arctan\left[\frac{d_z}{D}\right] - \alpha_i$$

(5.100)

$$2\theta_f = \arctan\left[\frac{d_y}{D}\right]$$

(5.101)

where D is the distance between the sample and the detector and d_z and d_y are the vertical and horizontal distances, respectively, defined in Fig. 5.19.

The reason for which vector $\vec{q}$ is written in Eq. (5.99) as a function of angles α_i and α_f will become apparent in the following sections. Due to reflection and refraction effects by surfaces or interfaces, which are important for incidence (α_i) and emergence (α_f) angles close to the critical angle of total external reflection α_c, the vector $\vec{q} = \vec{k}_f - \vec{k}_i$ differs from the vector associated with scattering by nanoparticles. Furthermore, in classical (transmission) SAXS experiments, the angles of incidence, refraction, and emergence, with respect to the outer surfaces of the analyzed material, are exactly or approximately equal to 90 degrees; thus, the effects due to refraction are negligible in these cases.

3.1.3 Scattering Intensity due to Nanoparticles Deposited on the Substrate Surface

In the conventional SAXS geometry (in which the incident beam is normal to the sample surface), each point on the plane of detection of the scattered intensity is unequivocally associated with a certain scattering vector. For GISAXS measurements, this no longer holds true; in the most general case, the presence of interfaces generally causes a given position in the detector to simultaneously receive multiple intensities associated with different scattering vectors, as will be described ahead.

One example concerning nanoparticles deposited on a substrate is illustrated in Fig. 5.20. As shown in the figure, the detector is positioned to measure the intensity of the scattered beam passing through a slit, defining a direction that forms an angle α_f with the

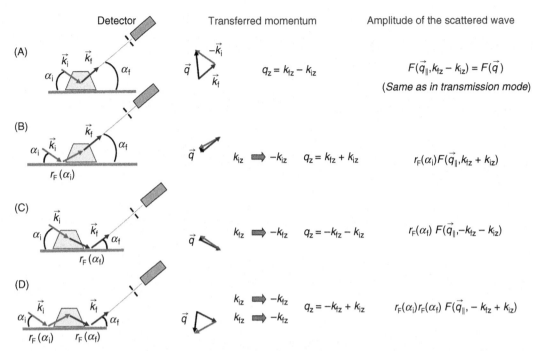

FIGURE 5.20 Representation of the different processes involved in X-ray scattering by a nanoparticle deposited on a flat substrate. (A) Incident beam scattered by a nanoparticle without interaction with substrate (as in conventional SAXS). (B) Incident beam reflected by the substrate before hitting the nanoparticle. (C) Beam scattered by the nanoparticle and then reflected by the substrate. (D) Both, incident and scattered beam waves reflected by the substrate. From Rauscher et al. (1999).

sample surface. Note that the four possibilities shown in Fig. 5.20 lead to scattered beams hitting the detector in the same point (Rauscher et al., 1999).

In the case illustrated in Fig. 5.20A, the incident beam is scattered by the nanostructure without interaction with the substrate. Here, the vector $\vec{k}_i$ has the same direction as the incident beam. The vertical component of the scattering vector can be simply written as $q_z = k_{fz} - k_{iz}$, similar to conventional SAXS transmission geometry. Fig. 5.20B shows the case of the incident beam reflecting on the substrate, inverting the sign of the vertical component of the incident wave vector prior to its interaction with the particle. The vertical component of the scattering vector is given by $q_z = k_{fz} + k_{iz}$. In Fig. 5.20C, the beam scattered by the sample is reflected by the substrate, yielding $q_z = -k_{fz} - k_{iz}$. The situation illustrated in Fig. 5.20D shows that both the incident and nanoparticle-scattered beams are reflected by the substrate, and, thus, $q_z = -k_{fz} + k_{iz}$. Note that in the situations illustrated in Fig. 5.20 involving reflection by the substrate, only the scattering vector component perpendicular to the sample surface, q_z, is influenced; the other components remain unaffected (Rauscher et al., 1999).

The probability of each event illustrated in Fig. 5.20A–D is proportional to the Fresnel reflection coefficients of the incident and scattered beams, respectively, given by

$$r_F(\alpha_i) = \frac{k_{iz} - k_{tz}}{k_{iz} + k_{tz}}$$

$$(5.102)$$

and

$$r_F(\alpha_f) = \frac{k_{fz} - k_{tz}}{k_{fz} + k_{tz}} \tag{5.103}$$

where $k_{fz} = -k \sin \alpha_f$ and k_{iz} and k_{tz} are defined according to Eqs. (5.93) and (5.94). Therefore, in addition to being a function of angles α_i and α_f, the probability of occurrence of each of the earlier processes is also dependent on the surface composition, radiation wavelength, and substrate surface roughness.

Thus, the total amplitude of the wave scattered by the particle is given by the sum of the waves associated with each of the various processes described, that is (Rauscher et al., 1999):

$$f(\vec{q}_\|, k_{iz}, k_{fz}) = F(\vec{q}_\|, k_{fz} - k_{iz}) + r_F(\alpha_i) F(\vec{q}_\|, k_{fz} + k_{iz}) +$$
$$r_F(\alpha_f) F(\vec{q}_\|, -k_{fz} - k_{iz}) + r_F(\alpha_i) r_F(\alpha_f) F(\vec{q}_\|, -k_{fz} + k_{iz}) \tag{5.104}$$

where $F(\vec{q}_\|, k_{iz}, k_{fz})$ is the function representing the particle's scattering amplitude, and $\vec{q}_\| = (q_x, q_y)$ is the component of vector $\vec{q}$ in the xy-plane.

The total scattering intensity is given by the product between $f(\vec{q}_\|, k_{iz}, k_{fz})$ and its complex conjugate.

The aforementioned example is not the only case in which reflection due to interfaces must be considered when calculating the GISAXS intensity. Other possibilities, which will not be discussed here, include, for example, nanoparticles embedded in layers or multilayers with reflective interfaces. In many of these cases, to obtain reliable results, it is necessary to measure the reflectivity rather than simply obtaining it as a modeling parameter. Cases involving more complex geometries are discussed by Lazzari (2002).

3.1.4 Nanoparticles Buried Below the Surface of a Given Substrate

As mentioned previously, for the case in which the nanoparticles are buried in the substrate, the incident angle must be adjusted so that the X-ray beam can penetrate to the depth at which they are located. If this depth and the refractive index are known, an appropriate incidence angle can be calculated using Eq. (5.90). If the parameters of this equation are not known, the value of α_i that maximizes the scattered intensity can be obtained experimentally by comparing GISAXS intensities measured at different values of α_i.

Because X-rays undergo refraction when crossing the air/sample interface, the scattering vector $\vec{q}$, determined from the incident beam and the beam scattered outside the sample, does not correspond to the scattering vector $\vec{\tilde{q}}$ within the sample. This occurs because of the following reasons: (1) the X-ray beam that hits the nanoparticles within the sample is the refracted beam, whose wave vector $\vec{k}_t$ differs from $\vec{k}_i$ in direction and magnitude, and (2) by traversing the surface, the beams scattered by nanoparticles are also refracted. On the other hand, only the scattering vector component in the direction perpendicular to the surface—the z-axis—is affected. The value of that component within the sample is given by (Lazzari, 2002)

$$\tilde{q}_z = \tilde{k}_{fz} - k_{tz} \tag{5.105}$$

where $\tilde{k}_{fz}$ is the z scalar component of the scattered beam wave vector within the sample (substrate), given by

$$\tilde{k}_{fz} = \sqrt{n_s^2 k^2 - |k_{f\parallel}|^2} \tag{5.106}$$

where n_S is the refractive index of the substrate, $k_{f\parallel} = \sqrt{k_{fx}^2 + k_{fy}^2}$ is the component of the scattered beam wave vector parallel to the sample surface (xy plane), and k is the magnitude of the scattering vector outside of the sample. The directions of the wave vectors inside and outside of the sample are shown in Fig. 5.21. In addition to the change in direction, the refraction at the interface also modulates the intensity of the scattered beams. This modulation arises from the dependence of the Fresnel transmission coefficients on the angle formed by the incident and scattered beams at the interface. The Fresnel transmission coefficients of the incident and scattered beams are given by (Lazzari, 2002)

$$t_F(k_{iz}) = \frac{2k_{iz}}{k_{iz} + k_{tz}} \tag{5.107}$$

and

$$t_F(k_{fz}) = \frac{2k_{fz}}{k_{fz} + \tilde{k}_{fz}} \tag{5.108}$$

respectively, where k_{fz} is the z scalar component of the scattered beam wave vector outside of the sample.

Thus, the amplitude of the scattered wave is given by (Lazzari, 2002)

$$f(\tilde{q}_\parallel, \tilde{q}_z, \alpha_i, \alpha_f) = t_F(k_{iz}) t_F(k_{fz}) F(\tilde{q}_\parallel, \tilde{q}_z) \tag{5.109}$$

The function $t_F(k_{fz})$ leads to an intensity maximum for the scattered beams whose angles with the sample surface are approximately equal to α_c. This scattered intensity maximum is known as the Yoneda peak (Yoneda, 1963).

Another factor that modulates intensity is the beam attenuation during its path in the sample. As shown in Fig. 5.21A, the length traveled by the scattered beams in the sample decreases as α_f increases. For typical scattering angles in GISAXS experiments, the dependence of attenuation on α_f is given by (Cullity, 1956)

$$A = \exp\left[\frac{\mu d}{\sin \alpha_f'}\right] \tag{5.110}$$

where α_f' is the angle between the scattered beam within the sample and the sample surface, μ is the linear absorption coefficient of the medium traversed by the beam, and d is the depth at which the nanoparticles are buried. α_f' can be determined from α_f and n_S by Snell's law: ($n_S \cos \alpha_f' = n_{air} \cos \alpha_f$).

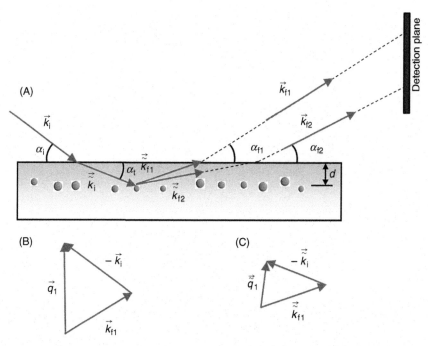

FIGURE 5.21 (A) Refraction of the incident and scattered beams after traversing the air/sample interface. For clarity, only two scattered beams are represented. (B) The scattering vector $\vec{q}_1$ from beam 1 outside of the sample and (C) the scattering vector $\vec{\tilde{q}}_1$ corresponding to beam 1 inside the sample.

3.1.5 Examples of GISAXS Application

Studies applying the GISAXS technique in the structural characterization of nanoparticle-containing thin films have been widely published. The next section will present two such examples. The first concerns the study of a silica thin film doped with metallic Co atoms deposited on a silicon wafer, whereas the second corresponds to the characterization of multilayers formed by semiconducting lead telluride (PbTe) nanocrystals embedded in amorphous silica. Both studies were performed using the XRD2 beamline at the Brazilian Synchrotron Light Laboratory (LNLS), located in Campinas, Brazil, using an experimental setup specifically designed for GISAXS studies.

3.1.5.1 COBALT SILICIDE (CoSi$_2$) NANOPLATELETS BURIED IN MONOCRYSTALLINE SILICON

This section describes studies of nanostructures observed in silica thin films doped with Co atoms, deposited on monocrystalline Si wafers with different crystallographic orientations, followed by treatment at 700 and 750°C, using transmission electron microscopy (TEM) and GISAXS. The studies showed that thermal treatments promote the formation of spherical Co nanoparticles within the silica film and that Co atoms diffuse from the thin film into the monocrystalline Si, leading to the formation of CoSi$_2$ nanoplatelets a few atomic layers within the Si.

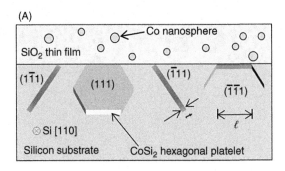

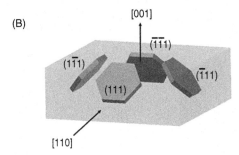

FIGURE 5.22 Schematic representation of the nanostructures observed by transmission electron microscopy. (A) Thin SiO$_2$ film containing spherical Co nanocrystals deposited on the (0 0 1) face of a monocrystalline Si wafer, inside of which hexagonal CoSi$_2$ nanoplatelets are formed; and (B) three-dimensional view of CoSi$_2$ nanohexagons, parallel to each of the four planes belonging to the Si crystallographic form {1 1 1}. *Part B extracted from Kellermann G, Montoro LA, Giovanetti LJ, Santos Claro PC, Zhang L, Ramirez AJ, Requejo FG, Craievich AF: Formation of an extended CoSi$_2$ thin nanohexagons array coherently buried in silicon single crystal, Appl Phys Lett 100:063116, 2012. Reprinted with permission from AIP Publishing LLC.*

TEM images showed that the nanoplatelets are quasiregular hexagons that are relatively uniform in size. The crystal lattice of the CoSi$_2$ nanoplatelets is consistent with the Si lattice, with the largest facet of the nanoplatelets parallel to one of the four planes of the Si {1 1 1} crystallographic form. A schematic representation of the case in which the surface of the Si substrate is perpendicular to the Si[0 0 1] crystallographic orientation is shown in Fig. 5.22.

To obtain structural information about a large surface area, the same sample was also studied by GISAXS. Fig. 5.23A shows the GISAXS intensity obtained experimentally. Taking into account the refraction effect of X-rays on the sample surface, the total GISAXS intensity for the case in which the Si substrate surface is parallel to the Si(0 0 1) planes is given by (Kellermann et al., 2012)

$$I \propto |t_F(\alpha_i)|^2 |t_F(\alpha_f)|^2 \left(c_R \sum_{hkl} | A_{\text{hex}(hkl)}(\alpha_i, \phi, q_x, q_y, \tilde{q}_z, \ell, t)|^2 + \right.$$

$$\left. \int | A_{\text{sph}}(q_x, q_y, \tilde{q}_z, R)|^2 \, N_{\text{sph}}(R) dR \right)$$

(5.111)

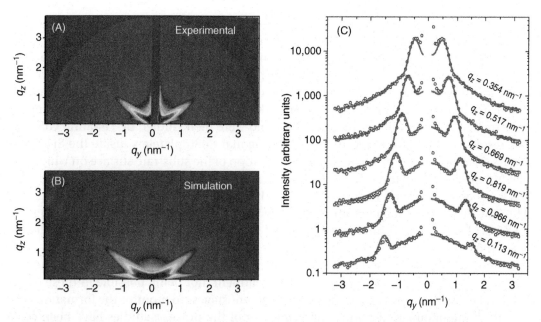

FIGURE 5.23 (A) Experimental GISAXS intensity measured with a 2D position-sensitive X-ray detector; (B) intensity calculated using the parameters derived from a best-fit procedure; (C) experimental (*circles*) and calculated (*lines*) intensity profiles corresponding to the best fit using the model described in the text. *GISAXS*, Grazing-incidence small-angle X-ray scattering. *Part B extracted from Kellermann G, Montoro LA, Giovanetti LJ, Santos Claro PC, Zhang L, Ramirez AJ, Requejo FG, Craievich AF: Formation of an extended CoSi$_2$ thin nanohexagons array coherently buried in silicon single crystal,* Appl Phys Lett *100: 063116, 2012. Reprinted with permission from AIP Publishing LLC.*

where functions $t_F(\alpha_i)$ and $t_F(\alpha_f)$ are the Fresnel transmission coefficients of the incident and scattered beams, respectively, and c_R is the ratio between the number of hexagonal and spherical particles. $A_{sph}(q_x,q_y,\tilde{q}_z,R)$ is the scattering amplitude of a spherical particle with radius R, $N_{sph}(R)dR$ represents the number of spherical Co particles with radii between R and $R + dR$, and $A_{hex(hkl)}(\alpha_i,\phi,q_x,q_y,\tilde{q}_z,\ell,t)$ is the scattering amplitude of a hexagon with side length ℓ and thickness t. The indices $h\,k\,l$ in this function differentiate the four different orientations of the hexagons (Fig. 5.22), each being parallel to one of the four planes of the Si {1 1 1} crystallographic form; α_i corresponds to the angle between the incident beam and the Si surface, and ϕ is the azimuthal angle.

Fig. 5.23C shows the $I \times q_y$ intensity profiles for different values of q_z. The continuous curves correspond to the best fit, assuming that the total intensity is given by the sum of the intensities due to (1) a monodisperse set of regular CoSi$_2$ nanohexagons buried within the Si monocrystal and parallel to the Si {1 1 1} crystallographic form and (2) Co nanospheres with a radius distribution function $N_{sph}(R)$ given by a lognormal function.

For comparison with the experimental intensity, Fig. 5.23B shows the intensity $I(q_y, q_z)$ calculated using the parameters obtained from the best-fit procedure using Eq. (5.111). This procedure allowed a determination of the thickness and lateral dimension of the hexagons, $t = 2.5$ nm and $\ell = 19.5$ nm, respectively; these values are in excellent agreement with those calculated from TEM images of the same sample.

To evaluate the effect of the crystallographic orientation of the Si substrate surface in the formation of nanohexagons, the same method was employed in the study of thin silica films doped with Co deposited on monocrystalline silicon wafers with surfaces parallel to the Si(0 1 1), Si(1 1 1), and Si(0 0 1) planes (Kellermann et al., 2015). The three samples were treated at 750°C for 1 h under He flow and were then studied by TEM and GISAXS. The results showed that the formation of hexagonal $CoSi_2$ platelets inside the Si matrix occurs regardless of the crystallographic orientation of the substrate surface on which the film is deposited. In all three cases, the crystallographic structure of the silicide nanoplatelets is coherently ordered to that of the Si, with the largest hexagon surface always parallel to the Si{1 1 1} crystallographic form. This result is attributed to the lower energy of the $CoSi_2${1 1 1}/Si{1 1 1} interface, compared to other interfaces in this system (Yalisove et al., 1989).

The GISAXS studies described earlier allowed for a determination of the nanostructure size and shape after the end of the thermal treatment, not being possible in these cases to characterize the structures in the early treatment stages. To monitor the formation and growth of the nanoparticles from the beginning of the process, studies have been conducted using a specifically modified camera to allow for GISAXS measurements during the thermal treatment. Such a camera was installed in the XRD2 beamline at LNLS, Campinas, Brazil. In that study, a SiO_2 film doped with Co deposited over Si(001) was investigated in situ during a isothermal treatment at 700°C (da Silva Costa et al., 2015).

The in situ studies using GISAXS showed that the formation of $CoSi_2$ platelets occurs in the first moments of treatment at 700°C. Their growth rate is higher in the first hour, after which it decreases; after 80 min of treatment, the nanoparticles cease to grow. The nanohexagon thickness and lateral size increase simultaneously, although the thickness increases at a lower rate. The volume fraction occupied by the $CoSi_2$ nanohexagons versus the duration of thermal treatment fits the behavior predicted by the Avrami equation for the case in which the number of supercritical nuclei is constant and one- and two-dimensional diffusion processes, related to the thickness growth and lateral growth of the nanohexagons, occur independently (Avrami, 1941).

It was also observed that spherical Co nanoparticles are present in the SiO_2 film before treatment at 700°C, indicating that they are formed during pretreatment at 500°C, used in the preparation of the Co-doped SiO_2 film. During treatment at 700°C, while the Co diffuses through the SiO_2/Si interface, which leads to the formation and growth of $CoSi_2$ nanohexagons, the volume fraction occupied by the Co nanoparticles in the film decreases. This finding was attributed to the dissolution of smaller Co nanoparticles formed at 500°C, which, at 700°C, have smaller radii than the critical radius.

3.1.5.2 MULTILAYERS OF PbTe NANOCRYSTALS IMMERSED IN SiO_2

Multilayers formed by PbTe nanocrystals immersed in SiO_2 over a Si monocrystal were prepared by alternately depositing PbTe by pulsed laser deposition (PLD) and amorphous

SiO_2 by plasma-enhanced chemical vapor deposition (PECVD) (Kellermann et al., 2010). Samples with different amounts of PbTe per bilayer were prepared by varying the number of laser pulses (50–200) used in the evaporation of a PbTe target. All samples studied had 20 bilayers and identical amounts of SiO_2 per bilayer.

Fig. 5.24A shows the GISAXS intensity for a sample prepared with 100 laser pulses. Horizontally elongated intensity maxima are observed in regular intervals along the direction normal to the sample surface. These intensity maxima were ascribed to constructive interference associated with the nearly equally spaced PbTe nanocrystal layers. Isotropic scattering that extends over a large angular range is observed, superimposed on these maxima. Similar features are observed in the GISAXS patterns for multilayers prepared with 150 and 200 laser pulses. However, the pattern for the sample prepared with 50 laser pulses per bilayer showed weak, approximately constant scattering intensity within the measured angular range, suggesting that, for that amount of deposited PbTe, the nanocrystals are much smaller than the minimum detectable size. Analysis of the experimental GISAXS patterns allowed for a determination of the average radius, the dispersion in radius values, and the degree of order in multilayers period.

Fig. 5.24B shows $I \times q_{\parallel}$ intensity profiles for different values of q_z, corresponding to the sample prepared using 100 laser pulses per bilayer. Solid lines represent the best fit to the experimental data, assuming that the distribution of nanocrystal radii can be described by a Gaussian function. The analysis of GISAXS curves indicates that with increasing amounts of deposited PbTe, (1) the average radius of the PbTe nanocrystals increases and (2) the total number of nanocrystals decreases. This finding suggests the action of a growth mechanism called coarsening (Lifshitz and Slyozov, 1961) and/or nanoparticle coalescence during the multilayer deposition process.

The paracrystal model interference function derived by Hosemann (1951) was used in the analysis of GISAXS intensity profiles along the direction normal to the sample surface. The average spacing between the PbTe nanocrystal layers and the dispersion around this average value were thus determined. The experimental and calculated profiles based on the described model are shown in Fig. 5.24C.

3.2 Anomalous or Resonant Small-Angle X-Ray Scattering

In general, it is not possible to distinguish between the contributions of different phases to the total intensity in SAXS experiments, where several nanostructured phases of different compositions may be present. For instance, in the case of catalysts comprising metallic nanoparticles dispersed in nanoporous substrates, the total SAXS intensity is given by the sum of contributions due to the pores and particles. In principle, it is not possible to distinguish between the two contributions. However, as described ahead, this difficulty can be overcome by performing several SAXS experiments using X-ray beams with different photon energies, far and close the absorption edges of the chemical elements in the sample. By varying the energy of the incident X-rays over values close to that of the absorption edge of a given element in the sample, the atomic scattering amplitude (and hence the

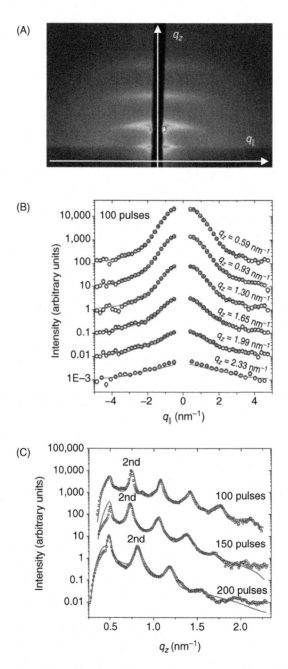

FIGURE 5.24 (A) GISAXS intensity corresponding to a PbTe/SiO₂ multilayer prepared by applying 100 laser pulses per layer; (B) horizontal intensity profiles and fitted curves obtained using the model described in the text; and (C) vertical intensity profiles for the multilayer and the corresponding best-fit curves obtained using Hosemann's paracrystal model. *GISAXS*, Grazing-incidence small-angle X-ray scattering. *Extracted from Kellermann G, Rodriguez E, Jimenez E, Cesar CL, Barbosa LC, Craievich AF: Structure of PbTe(SiO₂)/SiO₂ multilayers deposited on Si(111), J Appl Crystallogr 43:385–393, 2010. Reprinted with permission from IUCr.*

contribution of this chemical element to the total intensity) is changed. It can be shown that the variation in the scattered intensity provides information that reflects the concentration of the element in the different phases of the sample (Brumberger, 1993).

The anomalous small-angle X-ray scattering (ASAXS) technique makes use of the dependence of the atomic scattering factor with the photon energy to obtain selective structural information for a given chemical element. This technique plays an important role in the structural characterization of systems formed by phases with different compositions (Brumberger, 1993; Cromer and Liberman, 1981; James, 1965; Wendin, 1980). ASAXS finds important applications in materials science, particularly in systems such as metal alloys (Hoell et al., 2001; Hufnagel et al., 2001), catalysts (Haas et al., 2010; Haubold et al., 1997; Patel et al., 2004; Rasmussen et al., 2004), optical materials (Hoell et al., 2014), and polymers containing heavy atoms (Jokela et al., 2002; Nakanishi et al., 2016).

3.2.1 Atomic Scattering Factor

3.2.1.1 ATOMIC SCATTERING FACTOR FOR PHOTON ENERGIES FAR FROM THOSE CORRESPONDING TO THE ABSORPTION EDGES

For photon energies far larger than those corresponding to the absorption edges, the atomic scattering factor is given by the Fourier transform of the electron density function $\rho(\vec{r})$. For atoms with an isotropic electron density (Brumberger, 1993), we have

$$f_0(q) = \int_0^\infty 4\pi r^2 \rho(r) \frac{\sin qr}{qr} dr \tag{5.112}$$

Therefore, for $q = 0$ (i.e., the scattering angle $2\theta = 0$), f_0 equals the number of electrons, that is, the atomic number Z of the chemical element. As the scattering angle increases, the phase difference between waves scattered by different regions of the atom leads to a continuous reduction of f_0 for increasing values of $q = 4\pi \sin \theta/\lambda$. Plots of f_0 versus $\sin \theta/\lambda$ for X-ray photons with energies far from those corresponding to the atomic absorption edges can be calculated from the following equation (Cromer and Mann, 1968):

$$f_0\left(\frac{\sin\theta}{\lambda}\right) = \sum_{i=1}^4 a_i \exp\left[-b_i\left(\frac{\sin\theta}{\lambda}\right)^2\right] + c \tag{5.113}$$

where a_i, b_i, and c are tabulated parameters for each of the chemical elements (Wilson, 1995). Plots of f_0 versus $\sin \theta/\lambda$ for atoms with atomic number Z between 1 and 92 are shown in Fig. 5.25.

3.2.1.2 GENERAL DEFINITION OF THE ATOMIC SCATTERING FACTOR

For the more general case in which the energy of the X-ray photons is close to the energy of the absorption edge of the scattering atom, Eq. (5.112) is no longer valid, and the atom scatters the X-ray photons as if it had fewer than Z electrons. In this case, the atomic scattering factor is a complex number given by (Brumberger, 1993)

$$f_{\text{atm}}(\vec{q}, E) = f_0(\vec{q}) + f'(\vec{q}, E) + if''(\vec{q}, E) \tag{5.114}$$

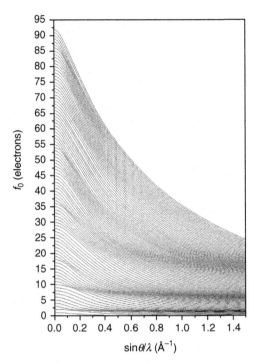

FIGURE 5.25 Atomic scattering factor (f_0) versus sin θ/λ for neutral atoms with atomic number Z between 1 and 92, calculated using the 9-parameter equation of Cromer and Mann (1968).

where f' and if'' are correction terms for the atom scattering factor. In addition to being dependent on $\bar{q}$, f' and f'' also depend on the photon energy E. The f' term, corresponding to the correction of the real part of f_{atm}, is a negative number and, as a consequence, it decreases the real part of the atomic scattering factor. In turn, f'' is related with X-ray absorption and fluorescence. As for f_0, the units of f' and f'' are given as the number of electrons; f' and f'' are related to the real and imaginary parts of the index of refraction according to Eqs. (5.88) and (5.89).

Fig. 5.26 shows the dependence of f' on photon energy for an isolated Bi atom (Henke et al., 1993). This plot shows that for photon energies far from the absorption edges, the absolute value of f' is small, and thus the atomic scattering factor $f_{atm}(E)$ is essentially equal to fo f_0. For energy values close to those of the absorption edges, it is necessary to precisely know the energy dependence of f_{atm} to accurately describe the scattering process.

For heavy elements, the K or L absorption edges are in the region of the electromagnetic spectrum corresponding to hard X-rays ($10 < E < 100$ keV). Because the L edge is associated to absorption processes involving six $2p$ electrons, instead of only two $1s$ electrons for K edge, the correction in the atomic scattering factor for energies close to the L edge is approximately three times greater than for the K edge. It is important to note that the wave functions of the electrons in the K and L energy levels are located close to the atomic nucleus. Consequently, the Fourier transform of these functions results in a contribution to f_{atm}

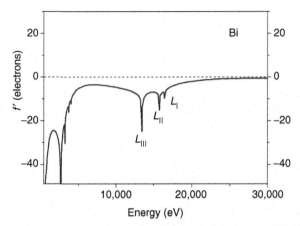

FIGURE 5.26 Real part of the correction term of the atomic scattering factor f' for an isolated Bi atom.

that is essentially independent of q. Because the scattering angles in the SAXS experiments are typically smaller than 5 degrees, f_0 does not differ significantly from Z. Thus, for SAXS experiments, the atomic scattering factor can be written as (Brumberger, 1993)

$$f_{atm}(E) = Z + f'(E) + if''(E) \tag{5.115}$$

The term f'' differs significantly from zero only for photon energies approximately equal to or larger than those corresponding to the absorption edges, where there is a significant increase in irradiated fluorescence. To minimize fluorescence, which would overlap with the scattered intensity, it is convenient for ASAXS experiments to be performed at photon energies slightly lower than the energies of the absorption edges (typically 5–10 eV below the edge).

3.2.2 SAXS Intensity for Photon Energies Close to Those of the Absorption Edges

When photons with energies close to those of the absorption edges are used, the effect introduced by the atomic scattering factor correction can be understood either as a reduction in the atom's ability to scatter radiation or as a reduction in the effective number of electrons participating in the scattering process.

As shown in the previous section, for small scattering angles, the SAXS intensity can be calculated from a function that describes the local average electron density. Thus, the atomic scattering factor per unit volume is given by (Brumberger, 1993)

$$\rho^{ef}(\vec{r}, E) = \frac{\sum_i n_i^a f_{atm}^i(E)}{\delta V(\vec{r})} \tag{5.116}$$

where n_i^a is the number of atoms of element i in the element with volume $\delta V(\vec{r})$ at position $\vec{r}$ and f_{atm}^i is the atomic scattering factor of atom i, which may depend on the photon energy E. The sum is made over all atoms in $\delta V(\vec{r})$.

Under these conditions, the scattering amplitude of a given object, labeled with the index j, can be written in the usual way (Brumberger, 1993), as follows:

$$F^j(\vec{q},E) = \int [\rho^{\text{ef}}(\vec{r},E) - \rho_0^{\text{ef}}(E)]\exp(-i\vec{q}\cdot\vec{r})dV \tag{5.117}$$

where the integral is calculated over the volume of the scattering object (nanoparticle) and ρ_0^{ef} is the average scattering factor per unit volume of the sample.

The intensity scattered by an arbitrary set of nanoobjects is given by the squared modulus of the sum of the amplitudes scattered by each of the objects times their respective phase factors, namely,

$$I(\vec{q},E) = \left| \sum_j F^j(\vec{q},E)\exp(-i\vec{q}\cdot\vec{R}_\parallel^j) \right|^2 \tag{5.118}$$

where $\vec{R}_\parallel^j$ is a vector describing the position of the nanoparticle's center of mass and $F^j(\vec{q},E)$ is the scattering amplitude of particle j.

3.2.3 Structural Analysis With Composition Selectivity

The following question can be posed: how can one take advantage of the dependence of the atomic scattering factor on the photon energy in SAXS studies involving systems composed of different chemical elements? To answer this question, we will examine the following hypothetical example.

A system is formed by two sets of spherical particles, each with a given composition. Let us consider the simple case in which one of these sets is formed by particles with radius R_A containing only the chemical element A, with the other set consisting of particles with radius R_B containing only element B. The particles are fully immersed in a medium that does not contain either element A or element B and whose electron density ρ_0 is smaller than the electron density of particles A and B. Suppose now that the SAXS intensity of this set of particles is measured with photon beams of two different energies, with at least one of these energies close to that of one of the absorption edges of element A: (1) a photon energy E_2, slightly lower than that corresponding to the edge, and (2) an energy E_1, a few hundred electron volts below the edge energy. The $f'(E)$ functions associated with elements A and B and the matrix are plotted in Fig. 5.27. Note that for energies close to that of the absorption edge of element A, there is significant variation in the f' factor of element A, whereas for the same photon energy interval, the f' functions corresponding to element B and the chemical elements in the matrix are almost invariant.

Assuming that this is a dilute system in which the particles are randomly distributed, the SAXS intensity is given by the sum of the individual intensities scattered by each particle (Glatter and Kratky, 1982):

$$\begin{aligned} I(q,E) = &\, N_A[\rho_A^{\text{ef}}(E) - \rho_0]^2[V(R_A)]^2 \,|\, F(q,R_A)|^2 \,+ \\ &\, N_B[\rho_B^{\text{ef}} - \rho_0]^2[V(R_B)]^2 \,|\, F(q,R_B)|^2 \end{aligned} \tag{5.119}$$

where the first term on the right corresponds to the contribution of particles composed of A atoms, denoted as $I_A(q, E)$, and the second term corresponds to the contribution of particles composed of B atoms, denoted as $I_B(q)$. N_A and N_B are the number of particles

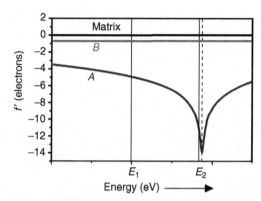

FIGURE 5.27 Schematic representation of the $f'(E)$ function for elements *A* and *B* and chemical elements in the matrix. The solid vertical lines indicate the energies E_1 and E_2 of the photon beams with which the SAXS intensity measurements were performed. The vertical dashed line indicates the energy of the absorption edge of element *A*. *SAXS,* Small-angle X-ray scattering.

consisting of elements *A* and *B*, respectively, and ρ_A^{ef} and ρ_B^{ef} are the corresponding atomic scattering factors per unit volume (or "effective electron densities"). $V(R_A)$ and $V(R_B)$ are the volumes, and $F(q, R_A)$ and $F(q, R_B)$ are the normalized form factors of the particles consisting of elements *A* and *B*, respectively. Note that in this case, only ρ_A^{ef} varies with photon energy, whereas ρ_B^{ef} and ρ_0 remain essentially constant within the energy range considered. Therefore, the difference between the SAXS intensities measured in E_1 and E_2 will be

$$I(q,E_1)-I(q,E_2)=[I_A(q,E_1)+I_B(q)]-[I_A(q,E_2)+I_B(q)] \tag{5.120}$$

Because the intensity $I_B(q)$ remains unchanged for photon energies E_1 and E_2, the difference in the contribution of particles formed by element *B* is canceled, thus yielding

$$I(q,E_1)-I(q,E_2)=I_A(q,E_1)-I_A(q,E_2) \tag{5.121}$$

Writing explicitly in terms of the shape factor, Eq. (5.121) becomes

$$\begin{aligned} I(q,E_1)-I(q,E_2)= N_A[V(R_A)]^2 \,|\, F(q,R_A)|^2 \times \\ \{[\rho_A^{ef}(E_1)-\rho_0]^2-[\rho_A^{ef}(E_2)-\rho_0]^2\} \end{aligned} \tag{5.122}$$

In other words, the difference between the SAXS intensities measured for these two photon energies, where at least one of these energies is close to the absorption edge of element *A*, leads to a function that is proportional to the intensity scattered by particles formed only of element *A*. The same procedure can be applied to the case in which set *B* are pores instead of particles.

3.2.4 Instrumental Aspects
Although the benefits of using ASAXS to characterize materials composed of nanostructures with different chemical elements are evident, several precautions involving the SAXS intensity measurements are required. It is essential that the instrumentation be specifically designed for this purpose. In this sense, an important boost to this technique was given by the development of sources that generate synchrotron radiation. The high intensity,

low angular divergence, and the broad spectrum of the radiation emitted by these sources provide obvious advantages over conventional X-ray instruments.

For light elements such as H, C, N, and O, the absorption edges appear at very low photon energies in the "soft X-ray" electromagnetic spectrum region (with energies less than ~1000 eV). Thus, for practical purposes, and due to the high absorption of low energy X-rays by matter, the ASAXS technique is most commonly applied to study materials that have elements with a relatively high atomic number, usually $Z > 20$. For this reason, ASAXS studies using photon energies close to those of light elements, such as those that constitute polymeric and biological materials, are difficult to perform.

3.2.4.1 CHOICE OF PHOTON ENERGY

As in the previous example, in most cases, it is desirable for the difference between the SAXS intensities measured using different photon energies to be as large as possible. Based on function $f'(E)$ shown in Figs. 5.26 and 5.27, it is clear that larger variations are obtained when photon energy values are chosen such that one is well below (~100 eV or more) and the other as close as possible to the corresponding absorption edge energy.

It is worth remembering at this point that to avoid excessive increases in absorption by the sample and in the emission of fluorescent radiation, all ASAXS measurements should be performed with photon energies lower than those of the absorption edges. To measure the SAXS intensity using photon energies as close as possible to those of the edges, while avoiding fluorescence, the radiation must have an adequate energy resolution. To obtain a proper contrast, the polychromatic radiation of the incident beam should be made monochromatic with high-resolution instruments, such that the ratio between the energy bandwidth ΔE (half-height width of the peak representing the function intensity vs. incident photon beam energy) and the average energy E_0 of the photons used in the experiments, $\Delta E/E_0$, is on the order of 10^{-4} or 10^{-3} at most. Taking into account the finite value of the energy bandwidth (usually a few electron volts), it is convenient for the ASAXS measurements to be performed with the highest energy being 5–15 eV lower than that corresponding to the absorption edge, depending on the resolution of the monochromator (Brumberger, 1993).

When multiple elements are present in the sample, SAXS measurements can be performed using photon energies close to those of the different edges, thus attaining more information and better result accuracy. In this case, however, the sample thickness should be adequately chosen such that the attenuation values are acceptable for all photon energies used.

3.2.4.2 DATA CORRECTION

3.2.4.2.1 NORMALIZATION OF SAXS INTENSITY AND SUBTRACTION OF PARASITIC SCATTERING Fig. 5.28 shows a typical setup scheme used in SAXS experiments. In the figure, two detectors are used to monitor I_0 and I_T, the intensities of the incident and transmitted beams, respectively. Thus, one can account for the continuous decrease in intensity emitted by the synchrotron source and differences in exposure times, as well as the dependence of the photon flux and sample attenuation on the X-ray photon energy.

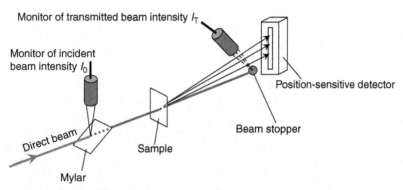

FIGURE 5.28 Schematic representation of the setup used to measure SAXS and the incident and transmitted beam intensities. *SAXS,* Small-angle X-ray scattering.

It is also necessary to consider the relationship between the efficiency of all detectors used and the photon energy. The relative efficiency of the detectors is generally determined from SAXS intensity measurements of a sample containing no elements with absorption edges within the energy range used in the experiment.

Finally, notice that the parasitic scattering, that is, the scattering intensity produced by windows, definition slits, and air gaps traversed by the beam, is superposed to the sample's SAXS scattering. The parasitic scattering is also a function of the photon energy. Therefore, for each SAXS measurement of a given sample with a given photon energy, it is also necessary to measure the parasitic scattering—without the sample—at the same photon energy. After intensities are corrected by the attenuation of the sample, the contribution due to parasitic scattering should be subtracted from the total intensity curve.

3.2.4.2.2 EXPERIMENTAL DETERMINATION OF THE REAL PART OF THE CORRECTION TERM OF THE ATOMIC SCATTERING FACTOR The precise knowledge of the atomic scattering factor, $f'(E)$, is fundamental in experiments aiming at determining the concentration of an element in the structure. On the other hand, tabulated or calculated atomic scattering factors are determined by assuming isolated atoms, under conditions in which the radiation is perfectly monochromatic. This assumption, however, does not correspond to reality, in which the atoms form chemical bonds. Under these conditions, the resonance energies corresponding to different observed absorption edges can differ by up to a few electron-volts from the energy of the absorption edges of the same atom when isolated. Moreover, in practice the energy bandwidth of the incident beam is not infinitely narrow but rather extends from a lower limit to an upper limit. For these reasons, to obtain accurate results, the function $f'(E)$ must be determined experimentally for each compound and for each specific condition employed in the experiment. The process for the experimental determination of $f'(E)$ is described in the subsequent text.

In the first stage, the X-ray transmission function is measured, that is, the ratio between the transmitted and incident intensity $T_r = I_T / I_0$, as a function of photon energy E. This step is done in a certain energy range, around the region corresponding to the absorption edge (typically between 500 eV below and above the edge). The linear absorption coefficient can then be calculated using the following relationship (Cullity, 1956):

$$\mu(E) = -\frac{\ln T_r}{t_a} \tag{5.123}$$

where t_a is the sample thickness.

Meanwhile, $\mu(E)$ is related to $f''(E)$—the function describing the imaginary part of the atomic scattering factor—through the following equation (James, 1965):

$$f''(E) = \frac{2\pi\varepsilon_0}{\rho_a} \frac{Emc}{he^2} \mu(E) \tag{5.124}$$

where m and e are the mass and charge of the electron, respectively, c is the speed of light in vacuum, h is the Planck constant, ε_0 is the electric permittivity of vacuum, and ρ_a is the number of atoms per unit volume.

The next step involves replacing the experimentally determined $f''(E)$ interval in $f_T''(E)$, which represents the scattering factor of the isolated atom (Henke et al., 1993).

Finally, $f'(E)$ can be calculated from $f''(E)$ using the Kramers–Kronig dispersion relation (Kramers, 1927; Kronig, 1926):

$$f'(E) = \frac{2}{\pi} \int_0^\infty \frac{E'f''(E')}{E'^2 - E^2} dE' \tag{5.125}$$

3.2.5 Application Example

An application of the ASAXS technique in the study of a sodium-borate glass ($72B_2O_3$–$28Na_2O$) containing a dilute set of spherical and homogeneous Bi nanocrystals randomly dispersed in a glass matrix is described in the subsequent text. A priori, the matrix could contain minor electron density inhomogeneities due to statistical fluctuations and, occasionally, due to the presence of nanopores and particles formed from the crystallization of the glass. Under these conditions, the total SAXS intensity is given by (Glatter and Kratky, 1982)

$$I(q,E) = [\rho_{Bi}(E) - \rho_0]^2 \left(\frac{4\pi}{3}\right)^2 \int_{R_{min}}^{R_{max}} |F(q,R)|^2 \, N(R)R^6 \, dR + I_v(q) \tag{5.126}$$

wherein

$$F(q,R) = 3\frac{\sin(qR) - qR\cos(qR)}{(qR)^3} \tag{5.127}$$

where $F(q, R)$ is the form factor of a homogeneous spherical particle with radius R, $N(R)dR$ is the number of nanoparticles with radii between R and $R + dR$, ρ_0 and ρ_{Bi} are the electron densities of the glass and Bi particles, respectively, and $I_v(q)$ is the intensity scattered by the glass. If the photon energy is close to that of one of the Bi absorption edges, its scattering factor per unit volume (or its "effective electron density"), ρ_{Bi}, will be a function of

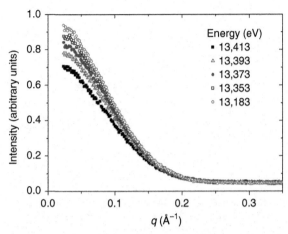

FIGURE 5.29 SAXS intensity curves versus *q* for 72B$_2$O$_3$–28Na$_2$O glass containing Bi nanocrystals, measured at the indicated energies. *SAXS*, Small-angle X-ray scattering.

energy E. In this same energy range, if the glass matrix contains no Bi atoms, the possible contribution to the total SAXS intensity due to statistical fluctuations in glass density, pores, and nanoparticles that may originate from the glass crystallization, $I_v(q)$, will be independent of the photon energy.

Fig. 5.29 shows SAXS intensity curves measured for different photon energies lower than that corresponding to the Bi absorption edge L$_3$ (13,419 eV). As expected, the intensity decreases as E approaches the edge. For high values of q, for which the scattering intensity due to Bi nanocrystals is smaller, the intensity becomes energy independent because the scattering factors of the atoms in the glass matrix are only slightly affected by the energy shift.

If the glass matrix contains no Bi atoms, the contribution of the glass to the total intensity remains the same, regardless of the energy. Thus, the SAXS intensity due solely to the contribution of Bi nanocrystals can be determined from the difference between the SAXS intensities measured at two different energies near the absorption edge of Bi, as the intensity due to the glass cancels out. A curve corresponding to the difference between the intensities measured at 13,183 and 13,413 eV is shown in Fig. 5.30. The same figure also shows the radius distribution function corresponding to the best fit to the experimental curve. The fitting procedure was carried out using the GNOM program (Semenyuk and Svergun, 1991).

The results of this ASAXS application example allow for the exclusive characterization of Bi-containing nanoparticles embedded in the glass matrix, that is, the results are independent of the presence of any electron density fluctuations arising from the glass, nanopores, or Bi-free glass nanocrystals. Similar ASAXS applications are used to determine the size distribution of metal nanoparticles embedded in nanoporous matrices, which are relevant models for the characterization of typical catalysts (Haas et al., 2010; Haubold et al., 1997).

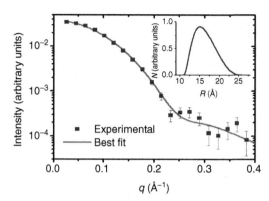

FIGURE 5.30 Square symbols: Difference between the SAXS intensities measured at 13,183 and 13,413 eV. Solid curve: Best-fit curve obtained using the GNOM program. The plot in the inset of the figure is the radius distribution function obtained by the best-fit procedure. *SAXS*, Small-angle X-ray scattering.

References

Abdala PM: *Materiales nanoestructurados y de grano fino de ZrO_2–Sc_2O_3 para celdas de combustible de óxido sólido de temperatura intermédia*, Doctorado en Ciencia y Tecnología, Mención Materiales, Instituto Sábato, Universidad Nacional de Gral, San Martín, Argentina, 2010.

Abdala PM, Craievich AF, Fantini MCA, Temperini MLA, Lamas DG: Metastable phase diagram of nanocrystalline ZrO_2–Sc_2O_3 solid solutions, *J Phys Chem C* 113:18661–18666, 2009.

Abdala PM, Fantini MCA, Craievich AF, Lamas DG: Crystallite size-dependent phases in nanocrystalline ZrO_2–Sc_2O_3, *Phys Chem Chem Phys* 12:2822–2829, 2010.

Abdala PM, Craievich AF, Lamas DG: Size-dependent phase transitions in nanostructured zirconia–scandia solid solutions, *RSC Adv* 2:5205–5213, 2012.

Acuña LM: *Conductores mixtos nanoestructurados para electrodos de celdas de combustible de óxido sólido de temperatura intermedia*, Tesis de Doctorado en Ciencias Físicas, Facultad de Ciencias Exactas y Naturales, Universidad Nacional de Buenos Aires, Buenos Aires, Argentina, 2012.

Acuña LM, Peña-Martínez J, Marrero-López D, Fuentes RO, Nuñez P, Lamas DG: Electrochemical performance of nanostructured $La_{0.6}Sr_{0.4}CoO_{3-\delta}$ and $Sm_{0.5}Sr_{0.5}CoO_{3-\delta}$ cathodes for IT-SOFCs, *J Power Sources* 196:9276–9283, 2011.

Als-Nielsen J, McMorrow D: *Elements of modern X-ray physics*, Chichester, 2011, John Wiley & Sons.

Avrami M: Granulation, phase change, and microstructure kinetics of phase change. III, *J Chem Phys* 9:177–184, 1941.

Borchert H: *Solar cells based on colloidal nanocrystals*, Cham, Heidelberg, New York, Dordrecht, London, 2014, Springer.

Brumberger H: *Modern aspects of small-angle X-ray scattering*, Dordrecht, 1993, Kluwer Academic Publisher.

Chacón P, Morán F, Díaz JF, Pantos E, Andreu JM: Low-resolution structures of proteins in solution retrieved from X-ray scattering with a genetic algorithm, *Biophys J* 74:2760–2775, 1998.

Chacón P, Díaz JF, Morán F, Andreu JM: Reconstruction of protein form with X-ray solution scattering and a genetic algorithm, *J Mol Biol* 299:1289–1302, 2000.

Chatterjee A, Pradhan SK, Datta A, De M, Chakravorty D: Stability of cubic phase in nanocrystalline ZrO_2, *J Mater Res* 9:263–265, 1994.

Clearfield A: Crystalline hydrous zirconia, *Inorg Chem* 3:146–148, 1964.

Cromer DT, Liberman DA: Anomalous dispersion calculations near to and on the long-wavelength side of an absorption edge, *Acta Crystallogr A* 37:267–268, 1981.

Cromer DT, Mann JB: X-ray scattering factors computed from numerical Hartree–Fock wave functions, *Acta Crystallogr A* 24:321–324, 1968.

Cullity BD: *Elements of X-ray diffraction*, Reading, MA, 1956, Addison-Wesley Publishing Co, Inc.

Da Silva Costa D, Huck-Iriart C, Kellermann G, Giovanetti LJ, Craievich AF, Requejo FG: In situ study of the endotaxial growth of hexagonal $CoSi_2$ nanoplatelets in Si(001), *Appl Phys Lett* 107:223101, 2015.

de Keijser ThH, Langford JI, Mittemeijer EJ, Vogels ABP: Use of the Voigt function in a single-line method for the analysis of X-ray diffraction line broadening, *J Appl Crystallogr* 15:308–314, 1982.

de Keijser ThH, Mittemeijer EJ, Rozendaal HCF: The determination of crystallite-size and lattice-strain parameters in conjunction with the profile-refinement method for the determination of crystal structures, *J Appl Crystallogr* 16:309–316, 1983.

Debye P: Zerstreuung von Rorntgenstrahlen, *Ann Phys* 46:809–823, 1915.

Ezquerra TA, Garcia-Gutierrez MC, Nogales A, Gomez M: *Applications of synchrotron light to scattering and diffraction in materials and life sciences*, Germany, 2009, Springer.

Fábregas IO: *Fases metaestables y nuevas propiedades de materiales nanoestructurados basados en ZrO_2. Aplicaciones en celdas de combustible de óxido sólido*, Tesis de Doctorado en Ciencias Químicas, Facultad de Ciencias Exactas y Naturales, Universidad Nacional de Buenos Aires, Buenos Aires, Argentina, 2008.

Fábregas IO, Lamas DG, Walsöe de Reca NE, Fantini MCA, Craievich AF, Prado RJ: Synchrotron X-ray powder diffraction and extended X-ray absorption fine structure spectroscopy studies on nanocrystalline ZrO_2–CaO solid solutions, *J Appl Crystallogr* 41:680–689, 2008.

Fendler JH: *Nanoparticles and nanostructured films*, Germany, 1998, John Wiley & Sons.

Fischer H, Oliveira Neto M, Napolitano HB, Craievich AF, Polikarpov I: The molecular weight of proteins in solution can be determined from a single SAXS measurement on a relative scale, *J Appl Crystallogr* 43:101–109, 2010.

Gaponenko SV: *Optical properties of semiconductor nanocrystals*, Australia, 1998, Cambridge University Press.

Garvie RC: The occurrence of metastable tetragonal zirconia as a crystallite size effect, *J Phys Chem* 69:1238–1243, 1965.

Giacovazzo C, Mónaco HL, Artioli G, Viterbo D, Milanesio M, Ferraris G, Gilli G, Gilli P, Zanotti G, Catti M: Fundamentals of crystallography. *International Union of Crystallography*, ed 3, New York, 2011, Oxford University Press.

Glatter O, Kratky O: *Small angle X-ray scattering*, London, 1982, Academic Press.

Guinier A, Fournet G: *Small angle scattering of X-rays*, New York, 1955, Wiley.

Haas S, Zehl G, Dorbandt I, Manke I, Bogdanoff P, Fiechter S, Hoell A: Direct accessing the nanostructure of carbon supported Ru–Se based catalysts by ASAXS, *J Phys Chem C* 114:22375–22384, 2010.

Haubold H-G, Wang XH, Goerigk G, Schilling W: In situ anomalous small-angle X-ray scattering investigation of carbon-supported electrocatalysts, *J Appl Crystallogr* 30:653–658, 1997.

Henke BL, Gullikson EM, Davis JC: X-ray interactions: photoabsorption, scattering, transmission, and reflection at E=50–30000 eV, Z=1–92, *Atomic Data Nucl Data Tables* 54:181–342, 1993.

Hoell A, Bley F, Wiedenmann A, Simon JP, Mazuelas A, Boesecke P: Composition fluctuations in the demixed supercooled liquid state of $Zr_{41}Ti14Cu_{12.5}Ni_{10}Be_{22.5}$: a combined ASAXS and SANS study, *Scr Mater* 44:2335–2339, 2001.

Hoell A, Varga Z, Raghuwanshi VS, Krumrey M, Bocker C, Rüssel C: ASAXS study of CaF2 nanoparticles embedded in a silicate glass matrix, *J Appl Crystallogr* 47:60–66, 2014.

Hosemann R: Die parakristalline feinstruktur natürlicher und synthetischer eiweisse. visuelles näherungsverfahren zur bestimmung der schwankungstensoren von gitterzellen, *Acta Crystallogr* 4:520–530, 1951.

Hufnagel TC, Gu XF, Munkholm A: Anomalous small-angle X-ray scattering studies of phase separation in bulk amorphous Zr52.5Ti5Cu17.9Ni14.6Al10, *Mater Trans* 42:562–564, 2001.

James RW: *The optical principles of the diffraction of X-rays*, Ithaca, 1965, Cornell University Press.

Jokela K, Serimaa R, Torkkeli M, Etelaniemi V, Ekman K: Structure of the grafted polyethylene-based palladium catalysts: WAXS and ASAXS study, *Chem Mater* 14:5069–5074, 2002.

Juárez RE, Lamas DG, Lascalea GE, Walsöe de Reca NE: Synthesis and structural properties of zirconia-based nanocrystalline powders and fine-grained ceramics, *Defect Diffusion Forum* 177–178:1–26, 1999.

Juárez RE, Lamas DG, Lascalea GE, Walsöe de Reca NE: Synthesis of nanocrystalline zirconia powders for TZP ceramics by a nitrate-citrate combustion route, *J Eur Ceram Soc* 20:133–138, 2000.

Kasai H, Escaño MCS: *Physics of surface, interface and cluster catalysis*, Bristol, UK, 2016, IOP.

Kellermann G, Vicentin F, Tamura E, Rocha F, Tolentino H, Barbosa A, Craievich AF, Torriani I: The small angle X-ray scattering beamline at the Brazilian synchrotron light laboratory, *J Appl Crystallogr* 30:880–883, 1997.

Kellermann G, Rodriguez E, Jimenez E, Cesar CL, Barbosa LC, Craievich AF: Structure of PbTe(SiO$_2$)/SiO$_2$ multilayers deposited on Si(111), *J Appl Crystallogr* 43:385–393, 2010.

Kellermann G, Montoro LA, Giovanetti LJ, Santos Claro PC, Zhang L, Ramirez AJ, Requejo FG, Craievich AF: Formation of an extended CoSi$_2$ thin nanohexagons array coherently buried in silicon single Crystal, *Appl Phys Lett* 100:063116, 2012.

Kellermann G, Montoro LA, Giovanetti LJ, Santos Claro PC, Zhang L, Ramirez AJ, Requejo FG, Craievich AF: Controlled growth of extended arrays of CoSi$_2$ hexagonal nanoplatelets buried in Si(001), Si(011) and Si(111) wafers, *Phys Chem Chem Phys* 17:4945, 2015.

Kirkpatrick S, Gelatt CD Jr, Vecchi MP: Optimization by simulated annealing, *Science* 220:671–680, 1983.

Klimov VI: *Semiconductor and metal nanocrystals: synthesis and electronic and optical properties*, Oxfordshire, 2005, Taylor & Francis e-Library.

Klug HP, Alexander LE: *X-ray diffraction procedures for polycrystalline and amorphous materials*, New York, 1974, John Wiley & Sons.

Kramers HA: La Diffusion de la Lumière par les Atomes, *Atti Cong Intern Fisica (Transactions of Volta Centenary Congress)* Como 2:545–557, 1927.

Kronig RL: On the theory of the dispersion of X-rays, *J Opt Soc Am* 12:547–557, 1926.

Lamas DG: *Estudio de las interfaces óxido de conducción iónica/óxido semiconductor/gas—Obtención de un nuevo sensor de gases*, Tesis de Doctorado en Ciencias. Físicas, Facultad de Ciencias Exactas y Naturales, Universidad Nacional de Buenos Aires, Buenos Aires, Argentina, 1999.

Lamas DG, Fuentes RO, Fábregas IO, Fernández de Rapp ME, Lascalea GE, Casanova JR, Walsöe de Reca NE, Craievich AF: Synchrotron X-ray diffraction study of the tetragonal-cubic phase boundary of nanocrystalline ZrO$_2$–CeO$_2$ synthesized by a gel-combustion process, *J Appl Crystallogr* 38:867–873, 2005.

Lamas DG, Rosso AM, Suarez Anzorena M, Fernández A, Bellino MG, Cabezas MD, Walsöe de Reca NE, Craievich AF: Crystal structure of pure ZrO$_2$ nanopowders, *Scr Mater* 55:553–556, 2006.

Lambert RM, Pacchioni G: *Chemisorption and reactivity on supported clusters and thin films: towards an understanding of microscopic processes in catalysis*, Netherlands, 1997, Kluwer Academics Publishers.

Langford JI: A rapid method for analysing the breadths of diffraction and spectral lines using the Voigt function, *J Appl Crystallogr* 11:10–14, 1978.

Langford JI, Wilson AJC: Scherrer after sixty years: a survey and some new results in the determination of the crystallite size, *J Appl Crystallogr* 11:102–113, 1978.

Langford JI, Delhez R, de Keijser ThH, Mittemeijer EJ: Profile analysis for microcrystalline properties by the Fourier and other methods, *Aust J Phys* 41:173–187, 1988.

Lazzari R: IsGISAXS: a program for grazing-incidence small-angle X-ray scattering analysis of supported islands, *J Appl Crystallogr* 35:406–421, 2002.

Liakakos N, Achkar C, Cormary B, Harmel J, Warot-Fonrose B, Snoeck E, Chaudret B, Respaud M, Soulantica K, Blon T: Oriented metallic nano-objects on crystalline surfaces by solution epitaxial growth, *ACS Nano* 9:9665–9677, 2015.

Lifshitz IM, Slyozov VV: The kinetics of precipitation from supersaturated solid solutions, *J Phys Chem Solids* 19:35–50, 1961.

Lintala M, Allahverdiyeva Y, Kidron H, Piippo M, Battchikova N, Suorsa M, Rintamäki E, Salminen TA, Aro EM, Mulo P: Structural and functional characterization of ferredoxin-NADP+ oxidoreductase using knock-out mutants of *Arabidopsis*, *Plant J* 49:1041–1052, 2007.

Mazdiyasni KS, Lynch CT, Smith JS: Metastable transitions of zirconium oxide obtained from decomposition of alkoxides, *J Am Ceram Soc* 49:286–287, 1966.

Mittemeijer EJ, Scardi P, editors: *Diffraction analysis of the microstructure of materials*, Berlin, Heidelberg, Alemanha, 2004, Springer-Verlag.

Mortier RM, Fox MF, Orszulik ST: *Chemistry and technology of lubricants*, ed 3, New York, 2010, Springer.

Nakanishi R, Machida G, Kinoshita M, Sakurai K, Akiba I: Anomalous small-angle X-ray scattering study on the spatial distribution of hydrophobic molecules in polymer micelles, *Polym J* 48:801–806, 2016.

Nascimento AS, Catalano-Dupuy DL, Bernardes A, Oliveira Neto M, Santos MAM, Ceccarelli EA, Polikarpov I: Crystal structures of *Leptospira interrogans* FAD-containing ferredoxin-NADP+ reductase and its complex with NADP+BMC, *Struct Biol* 7:69–81, 2007.

Nazarov A, Balestra F, Valeriya K, Flandre D: *Functional nanomaterials and devices for electronics, sensors and energy harvesting*, Cham, Heidelberg, New York, Dordrecht, London, 2014, Springer.

Patel M, Rosenfeldt S, Ballauff M, Dingenouts N, Pontoni D, Narayanan T: Analysis of the correlation of counterions to rod-like macroions by anomalous small-angle X-ray scattering, *Phys Chem Chem Phys* 6:2962–2967, 2004.

Patterson AL: A direct method for the determination of the components of interatomic distances in crystals, *Z Kristallographie A* 90:517–542, 1935.

Peng X, Mingos DMP: *Semiconductor nanocrystal and silicate nanoparticles*, Berlin, Heidelberg, Germany, 2005, Springer-Verlag.

Peng X, Schlamp MC, Kadavanich AV, Alivisatos AP: Epitaxial growth of highly luminescent CdSe/CdS core/shell nanocrystals with photostability and electronic accessibility, *J Am Chem Soc* 119:7019–7029, 1997.

Petkov V: Nanostructure by high energy X-ray diffraction, *Mater Today* 11:28–38, 2008.

Pietsch U, Holy V, Baumbach T: *High-resolution X-ray scattering: from thin films to lateral nanostructures*, New York, 2004, Springer.

Rasmussen FB, Sehested J, Teunissen HT, Molenbroek AM, Clausen BS: Sintering of Ni/Al2O3 catalysts studied by anomalous small angle X-ray scattering, *Appl Catal A Gen* 267:165–173, 2004.

Rauscher M, Paniago R, Metzger H, Kovats Z, Domke J, Peisl J, Pfannes HD, Schulze J, Eisele I: Grazing incidence small-angle scattering from free standing nanostructures, *J Appl Phys* 86:6763–6769, 1999.

Roy S, Ghose J: Synthesis of stable nanocrystalline cubic zirconia, *Mater Res Bull* 35:1195–1203, 2000.

Scardi P, Leoni M, Lamas DG, Cabanillas ED: Grain size distribution of nanocrystalline systems, *Powder Diffraction* 20:353–358, 2005.

Semenyuk AV, Svergun DI: GNOM—a program package for small-angle scattering data processing, *J Appl Crystallogr* 24:537–540, 1991.

Squire PG, Himmel ME: Hydrodynamics and protein hydration, *Arch Biochem Biophys* 196:165–177, 1979.

Štefanić G, Popović S, Musić S: Influence of pH on the hydrothermal crystallization kinetics and crystal structure of ZrO_2, *Thermochim Acta* 303:31–39, 1997.

Stuhrmann HB: Ein neues verfahren zur bestimmung der oberfla chenform und der inneren struktur von gelosten globularen proteinen, *Z Physikalische Chemie Neue Folge* 72:177–198, 1970a.

Stuhrmann HB: Interpretation of small-angle scattering of dilute solutions and gases: a representation of the structures related to a one-particle scattering functions, *Acta Crystallogr A* 26:297–306, 1970b.

Svergun DI: Determination of the regularization parameter in indirect-transform methods using perceptual criteria, *J Appl Crystallogr* 25:495–503, 1992.

Svergun DI: Restoring low resolution structure of biological macromolecules from solution scattering using simulated annealing, *Biophys J* 76:2879–2886, 1999.

Svergun DI, Koch MHJ: Small-angle scattering studies of biological macromolecules in solution, *Rep Prog Phys* 66:1735–1782, 2003.

Svergun DI, Barberato C, Koch MHJ: Crysol—a program to evaluate X-ray solution scattering of biological macromolecules from atomic coordinates, *J Appl Crystallogr* 28:768–773, 1995.

Svergun DI, Petoukhov MV, Koch MHJ: Determination of domain structure of proteins from x-ray solution scattering, *Biophys J* 80:2946–2953, 2001.

Thompson P, Cox DE, Hastings JB: Rietveld refinement of Debye–Scherrer synchrotron X-ray data from Al_2O_3, *J Appl Crystallogr* 20:79–83, 1987.

Tolan M: *X-ray scattering from soft-matter thin films: materials science and basic research*, Berlin, Heidelberg, 1999, Springer-Verlag.

Voevodin AA, Shtansky DV, Levashov EA, Moore JJ: *Nanostructured thin films and nanodispersion strengthened coatings*, Netherlands, 2004, Kluwer Academics Publishers.

Volkova VV, Svergun DI: Uniqueness of *ab initio* shape determination in small-angle scattering, *J Appl Crystallogr* 36:860–864, 2003.

von Laue M: *Concerning the detection of X-ray interferences*, Nobel Lecture, 1915.

Warren BE: *X-ray diffraction*, New York, 1990, Dover Publications, Inc.

Wendin G: Anomalous X-ray scattering, *Physica Scripta* 21:535, 1980.

Williamson GK, Hall WH: X-ray line broadening from filed aluminium and wolfram, *Acta Metall* 1:22–33, 1953.

Wilson ACJ: *International tables for X-ray crystallography* (vol C), Boston, 1995, Kluwer Academic Press, pp 500–502.

Yalisove SM, Tung RT, Loretto D: Epitaxial orientation and morphology of thin $CoSi_2$ films grown on Si(100): effects of growth parameters, *J Vac Sci Technol A* 7:1472–1474, 1989.

Yoneda Y: Anomalous surface reflection of X-rays, *Phys Rev* 131:2010–2013, 1963.

The Rietveld method. In Young RA, editor: *International union of crystallography*, New York, 1995, Oxford University Press.

Zhang S: Nanostructured thin films and coating: functional properties vol 2, New York, 2010, CRC Press.

6

Surface Plasmon Resonance (SPR) for Sensors and Biosensors

Celina M. Miyazaki*, Flávio M. Shimizu**, Marystela Ferreira*

*FEDERAL UNIVERSITY OF SÃO CARLOS (UFSCAR), SOROCABA, SP, BRAZIL;
**SÃO CARLOS INSTITUTE OF PHYSICS, UNIVERSITY OF SÃO PAULO,
SÃO CARLOS, SP, BRAZIL

CHAPTER OUTLINE

1 Introduction

In 1902 Wood made the first observation of the surface plasmons being reported as anomalies in the light diffracted on a metallic diffraction grating. Later, Fano proved that these anomalies are connected to the excitation of electromagnetic surface waves on the surface of the diffraction grating (Homola, 2008). In 1968 Otto demonstrated that the drop in the reflected light intensity in the attenuated total reflection (ATR) method occurred because of the surface plasmon generation. In the same year, Kretschmann and Raether proposed another configuration of ATR mode for surface plasmon excitation. However, it was only in 1980 that the first demonstration of the surface plasmon resonance (SPR) as an optical sensor for the study of surface processes at metal surfaces for gas sensing occurred (Nylander et al., 1982). The marketing of a device based on the SPR occurred only in 1990 by BIAcore to detect interactions between proteins (Karlsson, 2004).

Since then, sensors and biosensors based on SPR principle have shown significant advances, becoming an important tool for characterizing and quantifying biomolecular interactions. SPR can be applied in many areas: life science, electrochemistry, gas phase, food, environment, medicine, etc. In the medical diagnosis field, the technique is useful for development of highly sensitive and specific biosensors, with fast response time, label-free methodologies with fewer biological samples, and laboratory-developed tests (Homola, 2008). In the field of food analysis, the presence of toxins, pathogens, drugs, and nutritional additives can be analyzed (Homola et al., 2002; Oh et al., 2004a). In the drug development field, SPR provides information about the interactions between the component and the target, accelerating the drug development process.

More recently development and studies on biosensors have attracted much attention because they allow real-time monitoring of interaction of different biomaterials, such as enzymes, growth factors, glycoproteins, nucleic acids, drugs, cells, and viruses, and even biospecific interactions such as those between antigen and antibody (Karlsson, 2004). Much information can be extracted with this system such as mechanisms of molecular interaction (binding), time of reaction and its dissociation (kinetics), ligand–analyte level (affinity), amount of analyte (concentration), and specificity, with the main advantage that this is a label-free method.

SPR exhibits some advantages over currently used conventional techniques, such as enzyme-linked immunosorbent assay (ELISA) and fluorescence analysis, that is, the analyte does not need to have specific properties (fluorescence, absorption or scattering bands) and does not require fluorescent or radioactive markers (Boozer et al., 2006; Homola, 2003), which facilitates the direct monitoring of interactions without additional steps or experimental protocols. SPR also allows the combination/coupling to other techniques, for example, mass spectroscopy (Lopez et al., 2003; Zalewska et al., 2009), SELEX (Dausse et al., 2016), and electrochemistry (Baba et al., 2010; Wang et al., 2010). Furthermore, miniaturization of SPR system has been studied, which will allow the use of portable systems as recently reported by Liu et al. (2015) who proposed a SPR device based on smartphone platform and its system achieved a limit of detection of 47.4 against 15.7 nM from a commercial SPR instrument for IgG detection; however, the cost difference is approximately 1000 times lower for the new system.

SPR technique is an optical method that uses the changes in the refractive index very close to the sensor surface caused by the binding between an analyte in solution and its ligand immobilized on the sensor surface (McDonnell, 2001). In other words, a special optical assembly allows the generation of the surface plasmons at a metal/dielectric interface in a specific condition (e.g., angle or wavelength of incident light). These parameters suffer shifts by the changes in the refractive index of the dielectric close to the interface caused by interactions between sensor-immobilized molecules and sample molecules. The process will be better discussed in Section 2.

In this chapter we discuss the surface plasmon generations, exemplifying the most common configurations. Hereafter, the basics involved in the SPR-based sensors are presented. Finally, we summarize some of the important applications in different areas.

2 Surface Plasmons

Usually optical excitation of surface plasmons is carried out by means of ATR method, as demonstrated by Otto and Kretschmann. Fig. 6.1 depicts a light beam going through a high-refractive-index medium (e.g., prism) to a low-refractive-index medium (e.g., glass); the light is partly reflected and refracted (*i* and *ii*), and up the incident light angle approaches θ_c (Snell's law, $\sin \theta_c = 1$) where the light is refracted away from the normal of the surface. When the incident angle θ is larger than θ_c, the light is completely reflected.

Meanwhile coating the glass surface at the interface with a noble metal thin film (e.g., gold, silver), the entire light will not be reflected and part of it will disappear into the metallic film. The angle, which is greater than the critical angle, that reaches the maximum of loss and the minimum intensity of reflected light (dip) is the SPR angle (θ_{SPR}). The excitations of the mobile electrons (plasmons) that oscillate at the surface of the metal film are called surface plasmons. When the wave vector of the incident light (momentum of photons) matches the wavelength of the surface plasmons, the electrons resonate, hence the term SPR.

The propagation constant of the surface plasmons is higher than that of the optical wave propagation in the dielectric and, thus, the surface plasmons cannot be excited directly by focusing an optical wave on a metal-covered dielectric. Using different ways, the propagation constant of the incident optical wave can be enhanced to match that of surface plasmons, for example, ATR by prism coupler (Gwon and Lee, 2010; Kretschmann and Raether, 1968; Otto, 1968) or optical waveguides (Dostálek et al., 2001), and by diffraction using a diffraction grating (Cullen et al., 1987; Lawrence et al., 1996; Ritchie et al., 1968).

The prism coupling shown in Fig. 6.2A is the most traditional. The light wave passes through the prism with high refraction index and it is totally reflected to the base of the prism creating an evanescent wave that propagates along the interface with

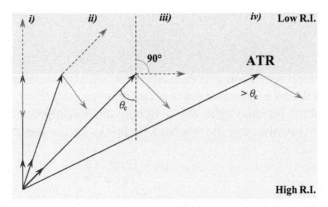

FIGURE 6.1 Schematic view of light incidence angle to obtain the ATR. *ATR,* Attenuated total reflection.

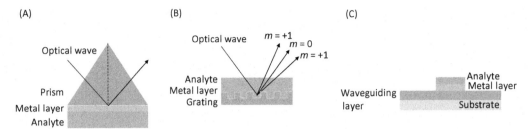

FIGURE 6.2 Widely used SPR configurations based on: (A) prism coupling, (B) grating coupling, and (C) optical waveguide coupling. *SPR,* Surface plasmon resonance.

the propagating constant that can be adjusted to match that of the surface plasmons (Daghestani and Day, 2010; Homola, 2003). The surface plasmons can also be excited by a diffraction grating, presented in Fig. 6.2B. In this case, the metal–dielectric interface is periodically distorted and the incident light is diffracted in various different angles. The diffracted beams have different momentum of the incident light that varies by multiples of the grating wave vector. In a specific angle of incidence, the diffracted light can couple to the plasmons when the momentum of the diffracted light parallel to the grating surface is equal to the propagation constant of the surface plasmons (Homola, 2008). Fig. 6.2C shows the optical waveguide configuration used for surface plasmon excitation. In a way similar to the ATR coupling, a light wave is guided by the waveguide to the metal interface and, in the case of phase matching, the light wave excites the surface plasmons in the outer interface of the metal (Homola et al., 1999).

Fig. 6.3 displays the schematic configuration of Kretschmann; in this case at the moment the wave vector of the incident light in the plane of the surface (k_s) matches the wave vector of the surface plasmon wave (SPW) in the metallic film (k_{sp}), solving Maxwell's equation a relationship can be expressed for resonance condition as follows (Gwon and Lee, 2010):

$$k_s = \frac{2\pi}{\lambda} \eta_p \sin\theta \tag{6.1}$$

$$k_{sp} = \frac{2\pi}{\lambda} \sqrt{\frac{\varepsilon_m \varepsilon_d}{\varepsilon_m + \varepsilon_d}} \tag{6.2}$$

where ε_d, ε_m, and η_p represent a complex dielectric constant of dielectrics, a complex dielectric constant of metal, and a refractive index of dense medium (prism), respectively, and λ is the wavelength of the incident light. The matching relationship, including the complex dielectric constant of metal ε_m and the resonant angle θ_{SPR}, can be described as follows:

$$k_{sp} = k_s, \quad \theta_{SPR} = \sin^{-1} \sqrt{\frac{\varepsilon_m \varepsilon_d}{\varepsilon_p(\varepsilon_m + \varepsilon_d)}} \tag{6.3}$$

because the complex refractive has the following correlation: $\eta = \sqrt{\varepsilon_p}$.

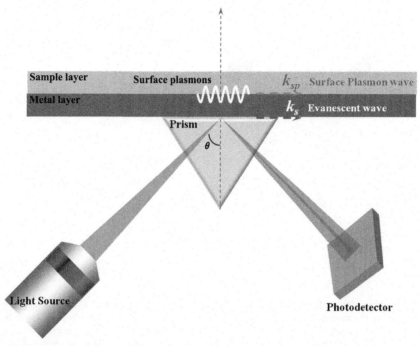

FIGURE 6.3 Schematic setup of Kretschmann configuration.

3 Surface Plasmon Resonance–Based Sensors

The propagation of the surface plasmon wave at the metal/dielectric interface is very sensitive to variations in the refractive index of the dielectric properties of its background (Dostálek et al., 2001; Homola et al., 1999), which enables the SPR application to investigate biochemical reactions that occurs close to the metal surface. Based on the characteristic of the light that is to be measured, the SPR sensor can be classified into angular-, wavelength-, or intensity-modulated systems. In the most common and sensitive method, the angular scanning, a monochromatic light is directed in different angles of incidence to the metal surface and the excitation of plasmons causes the absorption of light and it is observed as a dip in the angular spectrum of the reflected light (Löfås et al., 1991). In wavelength modulation SPR the surface plasmons are excited by a collimated light wave with multiple wavelengths and the plasmon excitation is observed as a minimum in the wavelength spectrum of the reflected light (Piliarik and Homola, 2009). In the intensity modulation SPR sensor the intensity of the light in a specific angle of incidence and wavelength is monitored. Although the performance using the intensity modulation is usually lower, this mode is widely used, in particular for SPR imaging techniques (Scarano et al., 2010).

For explaining the sensing applications of SPR, we discuss in more detail the most widely used configuration: the prism coupling in Kretschmann configuration. The light is

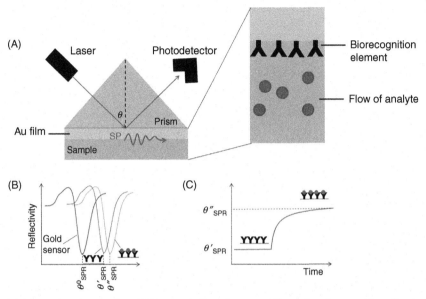

FIGURE 6.4 (A) Schematic representation of SPR setup in Kretschmann configuration. At a specific angle, the light energy is transferred to the surface plasmons causing a minimum intensity in the angular scanning, as can be seen in (B). The minimum reflectivity position, θ_{SPR}, changes after antibody immobilization on sensor surface (θ'_{SPR}) and also after antibody–antigen interaction (θ''_{SPR}). (C) The sensorgram (SPR angle variation as a function of time) allows the real-time monitoring of the θ_{SPR} during the antibody–antigen interaction. *SPR*, Surface plasmon resonance.

totally reflected at the base of the prism and surface plasmons are generated at the metal/dielectric interface (Homola, 2008). In a specific angle of incidence θ_{SPR}, part of light energy is transferred to the surface plasmons that propagates along the metal film surface as described in Fig. 6.4A (Homola, 2008) and, consequently, a minimum of reflectance is observed in the angular scanning (Löfås et al., 1991). The SPR angle strongly depends on the refraction index of surface, and then some change in the refractive index at the metal/dielectric interface results in intensity or position variation of the SPR minimum angle. The presence of adsorbed or linked molecules on the sensor surface shifts the refractive index, resulting in a change in the SPR angle (Boozer et al., 2006).

For the angular detection approach, it is possible to analyze the data in two different ways: angular scanning of incident light or by sensorgrams (monitoring the angular changes in function of time). Fig. 6.4B shows the angular scanning for three different situations: pure gold sensor exhibiting θ^0_{SPR}; after the antibody immobilization on surface, generating the sensor for a specific antigen, with the minimum reflectivity represented by θ'_{SPR}; and after detecting the target antigen, given by angular shift to θ''_{SPR}. Real-time detection analysis is usually performed by a sensorgram plot, which allows monitoring SPR angle variation as a function of time during the biomolecular interactions, as shown in Fig. 6.4C. The sensorgram allows to calculate (using an appropriate software) binding kinetic parameters as association/dissociation constants and affinity.

The SPR-based immunosensors especially have been widely explored because of their high sensitivity once the analyte concentration in biological systems is quite low. Usually, they have been developed for the measurement of antigens binding to antibody immobilized on the SPR sensor surface. It is useful for detecting analytes in complex biological media with specificity and sensitivity, with short detection time and simplicity (Oh et al., 2004a). Some strategies can also be performed to increase the SPR response against the analyte, for example, different antibody immobilization strategies, that is, random, covalent, and oriented using specific proteins, such as protein G (Oh et al., 2004a; Vashist et al., 2011). Additionally, sandwich assays by SPR (Cao et al., 2006; Homola et al., 2002) provide significant improvement in the SPR signal; however, they also increase steps to be performed in the analysis. Other researchers have performed studies using nanomaterials to enhance the performance of SPR analysis, for example, gold nanoparticles (Lee et al., 2006; Pieper-Fürst et al., 2005) and graphene (Singh et al., 2012, 2015).

One of the limitations of the SPR technique is the specificity of detection; if it is based on biomolecular recognition, some analyte molecules can bind to nontarget molecules on the sensor surface causing cross-sensitivity and generating a false-positive signal. Considering that, usually, the recognition biomolecules (e.g., antibodies) are attached to the sensor surface by bonding with a dextran or a thiolated acid layer (both presenting unreacted carboxyl groups), when the coverage is not effectively high, the remaining free carboxyl groups can attract the analyte (e.g., antigen) causing a false SPR response. Because of this, an important step of nonspecific interaction study is necessary to develop strategies to reduce or eliminate the nonspecific response. However, researchers have developed many effective strategies for overcome this drawback, for example, by addition of blocking molecules to deactivate the surface nonspecific groups (e.g., bovine serum albumin, ethylenediamine), changing buffer conditions (e.g., addition of detergents, salts), or using specific biotin–streptavidin approach. In this case, the target protein can be chemically bonded to biotin while streptavidin was attached to the sensor surface, guaranteeing a specific protein–ligand binding with high affinity (Morgan and Taylor, 1992).

4 SPR Applications

4.1 Materials Characterization

Although the main application of SPR is in the biomolecular analysis and biosensing, it is also useful for monitoring nanostructured film deposition, such as Langmuir–Blodgett (LB) (Çapan et al., 2010; Granqvist et al., 2013; Wagner and Roth, 1993) and layer-by-layer (LbL) (Advincula et al., 1996; Miyazaki et al., 2016; Pei et al., 2001) films. Therefore, angular scanning is carried out after each deposition step and by using Fresnel algorithm, it is possible to convert the variation of the θ_{SPR} to the thickness of the film adsorbed on the sensor surface.

The thickness d of nanostructured films such as LB and LbL is easily determined if the refractive index n is known or estimated. Fig. 6.5 shows the SPR angular curves for a pure

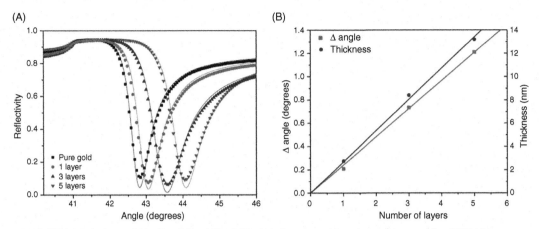

FIGURE 6.5 (A) Experimental (*dots*) and fitted (*lines*) SPR data for pure gold, and one, three, and five CdSA LB layers. (B) Angle and thickness variation as a function of the number of CdSA LB layers. *CdSA*, Cadmium stearic acid; *LB*, Langmuir–Blodgett; *SPR*, surface plasmon resonance. *Reprinted from Liang H, Miranto H, Granqvist N, Sadowski JW, Viitala T, Wang B, Yliperttula M: Surface plasmon resonance instrument as a refractometer for liquids and ultrathin films,* Sens Actuators B Chem *149:212, 2010.*

gold-covered glass sensor and covered with one, three, and five layers of cadmium stearic acid (CdSA) LB films (Liang et al., 2010). The successive layer deposition causes changes in the resonance angle for higher values as can be seen in Fig. 6.5A—measured curves (dots) were fitted by Winspall software (lines) based on Fresnel theory assuming the idealized layer model, where the layers are isotropic and the substrate is perfectly flat (Caruso et al., 1995). Assuming that $n = 1.45$, d can be obtained for each simulated layer, being equal to 2.68 nm for each CdSA monolayer (Liang et al., 2010). The linear relationship between n (and calculated d) and the number of CdSA layer is shown in Fig. 6.5B.

When n is unknown, the film thickness d can still be determined using two different approaches: acquiring SPR spectra in two different media or using different wavelengths of light source, allowing the determination of a unique value of d and n (Granqvist et al., 2013; Liang et al., 2010). For the different wavelength approach, two pairs of wavelength were used to perform the full angle scans in air, 655 and 782 nm, and 670 and 783 nm. Using the Winspall software, the relation between different n values and the corresponding d values was calculated. The d–n plots for different wavelengths for a LB monolayer of hydrogenated L-α-phosphatidylcholine (HSPC) are shown in Fig. 6.6A. A unique value of n and d was intended as an intersection point; however, as seen in the graphic, the two pairs of wavelength do not match at the same point, which can be justified by the fact that the measurements were performed in two different spots. The parameters d and n were determined by the average of the two cross-points in the d–n plots for the different laser pairs (Granqvist et al., 2013). From the different wavelength method n was determined as 1.334 and $d = 3.7$ nm for one layer of HSPC.

In the case of the two-medium analysis, d–n plot was obtained from simulation of SPR curves in two different media, air and water, in the same wavelength (785 nm). Fig. 6.6B

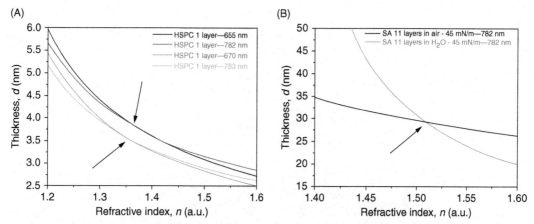

FIGURE 6.6 Examples of finding the intersection points from thickness *d* versus refractive index *n* of 1 monolayer of HSPC in air using different wavelength method (A) and 11 layers of SA using two-medium method (B). The arrows emphasize the intersection points. *HSPC*, Hydrogenated L-α-phosphatidylcholine; *SA*, stearic acid. *Reprinted from Granqvist N, Liang H, Laurila T, Sadowski J, Yliperttula M, Viitala T: Characterizing ultrathin and thick organic layers by surface plasmon resonance three-wavelength and waveguide mode analysis,* Langmuir *29:8561–8571, 2013.*

shows the *d–n* relation for stearic acid (SA). A unique value of *n* and *d* was obtained by the intersection point, being determined as *n* = 1.573 and *d* = 26.4 nm for 11 layers of SA film.

4.2 Medical Diagnosis

The SPR technology has contributed to development of biosensors with high sensitivity and reproducibility for medical diagnosis. Being a label-free technique and allowing the real-time interaction monitoring, the SPR biosensors have received special attention in the past years. Additionally, surface modifications and different approaches can be performed for sensitivity improvements, aiming a decrease in the limits of detection.

For example, cardiac troponin T (cTnT) is immediately released in the bloodstream during acute myocardial infarction (AMI) and it is a cardiospecific marker for myocardial cell damage (Dutra and Kubota, 2007). Dutra and Kubota developed a SPR sensor based on the biotinylated anti–cardiac troponin T. Fig. 6.7 shows the schematic representation of the biosensor that consists of a gold sensor covered with a dextran (carboxymethylated dextran) layer. Because of its carboxylic groups, dextran is coupled to the next streptavidin ligand layer via primary amino groups. Finally, biotinylated cTnT antibody is combined to the streptavidin forming the specific sensor for cTnT.

Fig. 6.8A shows the sensorgram of biotinylated anti-cTnT (black line) responding to the cTnT injections. In order to correct the experimental error or bias, a nonreagent against cTnT was used as control; in this case, the same concentration of goat antihuman IgG was injected (red line). In Fig. 6.6B the calibration curve for cTnT exhibits a plateau at ~7.0 ng/ mL and a linear range from 0.03 to 6.5 ng/mL, with detection limit of 0.01 ng/mL. The sensor was also tested in real samples and has detected cTnT in human serum without dilution with good specificity and reproducibility (Dutra and Kubota, 2007).

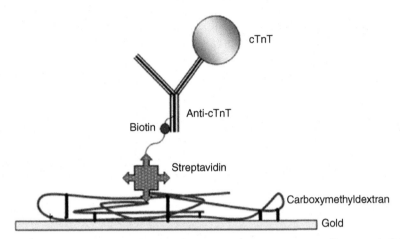

FIGURE 6.7 Schematic illustration for the sensor fabrication for cTnT detection. *cTnT*, Cardiac troponin T. *Reprinted from Dutra RF, Kubota LT: An SPR immunosensor for human cardiac troponin T using specific binding avidin to biotin at carboxymethyldextran-modified gold chip,* Clin Chim Acta *376:114–120, 2007.*

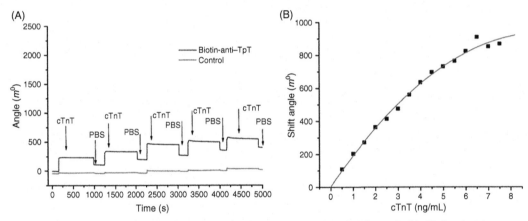

FIGURE 6.8 (A) Sensorgram of the successive injections of cTnT (0.2 ng/mL) in PBS. Channel 1 (black) showing the response of the biotinylated anti-TpT sensor and channel 2 showing the control antihuman IgG. (B) Calibration curve showing the relation between the SPR angle variation and the cTnT concentration. *cTnT*, Cardiac troponin T; *SPR*, surface plasmon resonance; *TnT*, troponin T. *Reprinted from Dutra RF, Kubota LT: An SPR immunosensor for human cardiac troponin T using specific binding avidin to biotin at carboxymethyldextran-modified gold chip,* Clin Chim Acta *376:114–120, 2007.*

Aiming to improve the detection of biomarkers for disease diagnostics, multiplexed systems, viz., surface plasmon resonance imaging (SPRI), have also been developed to allow the simultaneous detection of multiple DNA, RNA, and proteins at the same electrode that is integrated with microfluidic method. A schematic diagram of real-time SPRI experimental set-up for the simultaneous detection of biomolecules using microarrays is displayed in Fig. 6.9A and the integrated electrode microfluidic on Fig. 6.9B, in this system only localized sites that bioaffinity adsorption occurred will change the refractive index (Fasoli and Corn, 2015).

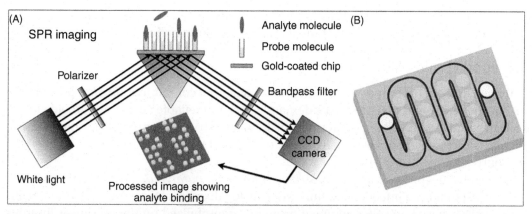

FIGURE 6.9 Schematic diagram of the SPRI experimental setup. (A) Bioaffinity adsorption onto a 45-nm gold thin film results in a local refractive index change. This leads to a difference in the monochromatic reflectivity image of a biopolymer microarray, which is obtained in the prism-coupled internal reflection geometry at the SPRI angle. (B) An example of a 16-element microfluidic SPRI chamber. The microfluidic channel has a cross-sectional area of 1 mm × 100 μm. The total cell volume is approximately 10 μL. *SPR*, Surface plasmon resonance; *SPRI*, surface plasmon resonance imaging. *Reprinted from Fasoli JB, Corn RM: Surface enzyme chemistries for ultrasensitive microarray biosensing with SPR imaging, Langmuir 31:9527–9536, 2015.*

Corn et al. developed a novel approach for the detection of miRNAs on locked nucleic acid (LNA) microarrays using the SPRI technique (Fang et al., 2006). First they modified electrode with LNA microarray through polyadenylation amplification (Fig. 6.10A), and later they detected three synthetic miRNA targets extracted from mouse liver: miR-16,

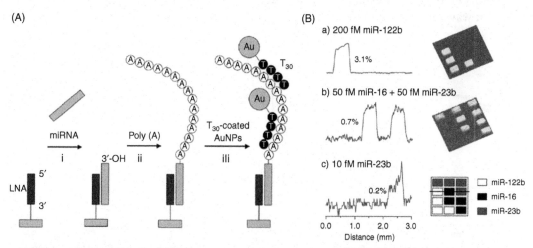

FIGURE 6.10 Surface poly(A)polymerase reactions for ultrasensitive miRNA detection. (A) Schematic of miRNA detection by polyadenylation and then coupling with nanoparticle-enhanced SPRI by the adsorption of T30 AuNPs to the poly(A) tail. (B) SPR difference images taken by drawing a line profile across the microarray image (*solid line*). *SPR*, Surface plasmon resonance; *SPRI*, surface plasmon resonance imaging. *Reprinted from Fang S, Lee HJ, Wark AW, Corn RM: Attomole microarray detection of microRNAs by nanoparticle-amplified SPR imaging measurements of surface polyadenylation reactions, J Am Chem Soc 128:14044–14046, 2006.*

miR-23b, and miR-122b (Fig. 6.10B); the microarray depicts a draw where each target was detected, achieving a limit of detection at attomolar concentration.

4.3 Food Quality Sensors

Pathogen detection, composition adulteration, bacterial contamination, and presence of toxins and environmental pollutants demand a fast and reliable analysis to provide information about food quality and safety for the food industry and control authorities. During the past years, biosensors especially based on the SPR technology have been applied for this purpose (Situ et al., 2010).

Oh et al. (2004b) developed a sensor for *Salmonella paratyphi*, a pathogenic microorganism in foodborne infections in humans, causing diarrhea, fever, and abdominal pain. The sensor is based on self-assembled protein G immobilized on the Au sensor in order to promote high orientation of the specific antibodies. For this, Au sensor is treated with protein G followed by monoclonal antibody against *S. paratyphi*. Then, the sensor was applied to detect *S. paratyphi*. Fig. 6.11A shows the SPR curves of Au sensor after being covered with protein G (curve a), after being covered with monoclonal antibody against *S. paratyphi* (curve b), and after reaction with *S. paratyphi* (curve c). As shown, the SPR angle shifted from 43.257 to 43.512 degrees by the immobilization of the antibody on the protein G–covered surface and from 43.512 to 44.142 degrees by its antigen binding to the sensor surface. Fig. 6.11B shows the relationship between the pathogen concentration and the SPR angle shift. The lowest limit of detection obtained by Oh et al. for this sensor was 10^2 CFU/mL, which was 4 orders of magnitude more sensitive than a standard ELISA.

Homola et al. (2002) developed a SPR sensor based on sandwich-type antibody assay based on wavelength modulation for *staphylococcal enterotoxin B* (SEB), which is 1 of a

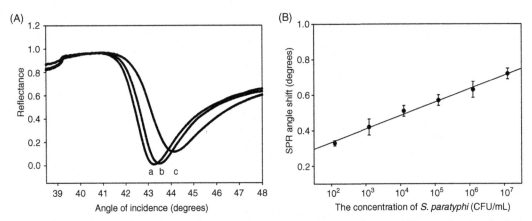

FIGURE 6.11 (A) Variation of SPR curves by adsorbing (a) self-assembled protein G layer and (b) monoclonal antibody against *S. paratyphi*, and (c) after detection of *S. paratyphi*. (B) Change in the SPR angle by binding between monoclonal antibody against *S. paratyphi* immobilized on the self-assembled protein G and various concentrations of *S. paratyphi*. SPR, Surface plasmon resonance. *Reprinted from Oh B-K, Lee W, Kim Y-K, Lee WH, Choi J-W: Surface plasmon resonance immunosensor using self-assembled protein G for the detection of* Salmonella paratyphi, *J Biotechnol 111:1–8, 2004.*

family of 10 major serological types of emetic enterotoxins produced by *Staphylococcus aureus*. In a *sandwich* assay format, the sample containing the analyte is brought in contact with the sensor covered by the capture antibodies. Then the sensor is incubated with a secondary antibody solution that binds to the previously captured antigen analytes, increasing the number of bound biomolecules and also the sensor response (Homola, 2003). The SPR substrates were functionalized by NHS–EDC for anchoring of anti-*SEB* IgG (capture antibody). Fig. 6.12 shows the response of the SPR sensor covered with the captured anti-SEB. After incubation in BSA–PBS, the SEB solution (analyte) was injected in the sample channel (response in Fig. 6.12A) while BSA–PBS was flowed in the reference channel as control (Fig. 6.12B). After washing the surface with BSA–PBS, the secondary antibody was injected into both the channels. As we can observe, the sample channel exhibits a strong response (approximately 4 nm) while the reference shows no significant response demonstrating good passivation of surface. Subtracting the sample channel response of the reference response we can have the corrected response in relation to the nonspecific effects. Fig. 6.12C shows the SPR response in a direct assay where the signal from SEB binding with the capture antibody is measured for each concentration. In Fig. 6.12D the response for the *sandwich* assay is shown; the signal measured is from the

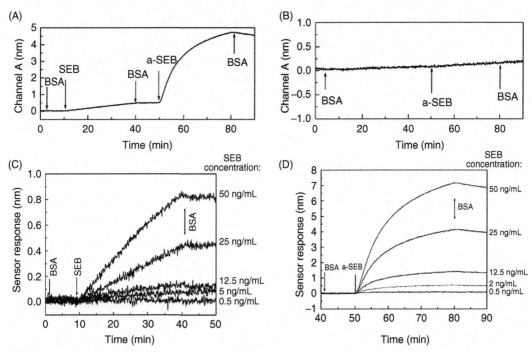

FIGURE 6.12 Sensorgram showing the response of *sandwich* assay sensor for SEB (25 ng/mL) in BSA–PBS solution (A) and the referencing channel signal (B). Response to different concentrations of SEB (direct assay) (C) and for the secondary antibodies (*sandwich* assay) (D). *SEB*, Staphylococcal enterotoxin B. *Reprinted from Homola J, Dostálek J, Chen S, Rasooly A, Jiang S, Yee SS: Spectral surface plasmon resonance biosensor for detection of staphylococcal enterotoxin B in milk,* Int J Food Microbiol *75:61–69, 2002.*

secondary antibodies by binding with the SEB already bonded in the sensor surface to the capture antibodies. The *sandwich* assay increases the sensor response by a factor of 10. The sensor was able to detect ~0.5 ng/mL of SEB in milk.

4.4 Drug Development

The SPR technique is a powerful tool for the pharmaceutical area. Sensorgram analysis gives information about binding level and rate of interactions, as association and dissociation constants. The determination of kinetic constants of biomolecular interactions is very important in the investigation of a biological system, allowing studies related to the selection and design of new molecules of therapeutic interest (Olaru et al., 2015). Also, determining the binding kinetics allows to establish the duration of action and clinical benefit of the chemical compound to a specific target. The main aims of the SPR measurements in the drug discovery are the following: (1) identify the binding typology of a pair of reactants; (2) determine the affinity constants related to the interaction processes occurring during the real-time analysis; (3) quantify association and dissociation rates; and (4) determine the concentration of the interacting partners.

For example, drug permeation studies have been done immobilizing liposomes on the SPR sensor mimetizing cell membranes. Kamimori et al. (2005) studied the affinity of two potential antimicrobial peptides for different compositions of lipid membranes: DMPC and DMPE. Fig. 6.13 shows the SPR sensorgrams showing the SPR angular changes when kalata B1 binds to the immobilized DMPC and DMPE on the sensor chip. A proportional increase in the sensor response with the peptide concentration indicates that the system has not reached saturation. Also, it is possible to conclude that a higher proportion of the kalata peptides binds to DMPE than to DMPC, confirmed by the affinity values (obtained by simulation of the sensorgrams using two-state reaction model) of 9.98×10^3 and 43.4×10^3 for the peptide–DMPC and peptide–DMPE, respectively (Kamimori et al., 2005).

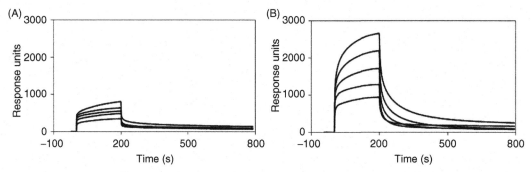

FIGURE 6.13 Sensorgrams for kalata B1 binding to (A) DMPC and (B) DMPE liposomes immobilized on the sensor chip surface (peptide concentration = 10–50 μM). *Reprinted from Kamimori H, Hall K, Craik DJ, Aguilar M-I: Studies on the membrane interactions of the cyclotides kalata B1 and kalata B6 on model membrane systems by surface plasmon resonance, Anal Biochem 337:149–153, 2005.*

The binding kinetic analysis showed that the two kalata peptides presented high affinity for phosphatidylethanolamine-containing membranes (e.g., DMPE), which suggests that the kalata peptides would bind preferentially to bacterial membranes (Kamimori et al., 2005). The authors attributed the selective binding of the peptides to DMPE to the different structure of the head group for each lipid, as DMPE contains the linear ethanolamine structure, while DMPC contains the bulky choline head group. Interaction parameters, such as association/dissociation and affinity constants, can be obtained by appropriate models, and software is available in the market, for example, TraceDrawer and Scrubber.

5 Final Remarks

The SPR technique has demonstrated huge advances since its first demonstration and it has been applied in several areas such as medical diagnosis, pharmaceutical area, material science, and food quality, among others. Being a label-free technique is one of the great advantages of SPR. Furthermore, high sensitivity and selectivity can be achieved by coating the sensor surface with specific biomolecules and different approaches can be used for preventing nonspecific bindings. It is well known that the SPR is a powerful tool for the development of new technologies in sensors and biosensors and there are many advances that are possible in the future regarding the SPR technique.

List of Symbols and Abbreviations

AMI	Acute myocardial infarction
ATR	Attenuated total reflection
BSA	Bovine serum albumin
CdSA	Cadmium stearic acid
cTnT	Cardiac troponin T
d	Thickness
DMPC	Dimyristoyl-L-α-phosphatidylcholine
DMPE	Dimyristoyl-L-α-phosphatidylethanolamine
DNA	Deoxyribonucleic acid
EDC	1-Ethyl-3-(3-dimethylaminopropyl)-carbodiimide
ELISA	Enzyme-linked immunosorbent assay
HSPC	Hydrogenated L-α-phosphatidylcholine
IgG	Immunoglobulin G
k_s	Wave vector of the incident light in the plane of the surface
k_{sp}	Wave vector of the surface plasmon wave in the metallic film
LB	Langmuir-Blodgett
LbL	Layer-by-layer
LNA	Locked nucleic acid
miRNA	Microribonucleic acid
n	Refractive index
NHS	*N*-hydroxysuccinimide
PBS	Phosphate buffered saline

RNA	Ribonucleic acid
SA	Stearic acid
SEB	Staphylococcal enterotoxin B
SELEX	Systematic evolution of ligands by the exponential enrichment
SPR	Surface plasmon resonance
SPRI	Surface plasmon resonance imaging
SPW	Surface plasmon wave
ε_d	Complex dielectric constant of dielectric
ε_m	Complex dielectric constant of metal
λ	Wavelength of the incident light
η_p	Refractive index of dense medium (prism)
θ_c	Critical angle
θ_{SPR}	Surface plasmon resonance angle
θ_{SPR}^0	Surface plasmon resonance angle for bare gold sensor
θ_{SPR}'	Surface plasmon resonance angle after antibody immobilization
θ_{SPR}''	Surface plasmon resonance angle after interaction with the analyte

References

Advincula R, Aust E, Meyer W, Knoll W: In situ investigations of polymer self-assembly solution adsorption by surface plasmon spectroscopy, *Langmuir* 12:3536–3540, 1996.

Baba A, Taranekar P, Ponnapati RR, Knoll W, Advincula RC: Electrochemical surface plasmon resonance and waveguide-enhanced glucose biosensing with *N*-alkylaminated polypyrrole/glucose oxidase multilayers, *ACS App. Mater Interfaces* 2:2347–2354, 2010.

Boozer C, Kim G, Cong S, Guan H, Londergan T: Looking towards label-free biomolecular interaction analysis in a high-throughput format: a review of new surface plasmon resonance technologies, *Curr Opin Biotechnol* 17:400–405, 2006.

Cao C, Kim JP, Kim BW, Chae H, Yoon HC, Yang SS, Sim SJ: A strategy for sensitivity and specificity enhancements in prostate specific antigen-α1-antichymotrypsin detection based on surface plasmon resonance, *Biosens Bioelectron* 21:2106–2113, 2006.

Çapan R, Özbek Z, Göktaş H, en S, nce FG, Özel ME, Stanciu GA, Davis F: Characterization of Langmuir–Blodgett films of a calix[8]arene and sensing properties towards volatile organic vapors, *Sens Actuators B Chem* 148:358–365, 2010.

Caruso F, Serizawa T, Furlong DN, Okahata Y: Quartz crystal microbalance and surface plasmon resonance study of surfactant adsorption onto gold and chromium oxide surfaces, *Langmuir* 11:1546–1552, 1995.

Cullen DC, Brown RGW, Lowe CR: Detection of immuno-complex formation via surface plasmon resonance on gold-coated diffraction gratings, *Biosensors* 3:211–225, 1987.

Daghestani HN, Day BW: Theory and applications of surface plasmon resonance, resonant mirror, resonant waveguide grating, and dual polarization interferometry biosensors, *Sensors* 10:9630–9646, 2010.

Dausse E, Barré A, Aimé A, Groppi A, Rico A, Ainali C, Salgado G, Palau W, Daguerre E, Nikolski M, Toulmé J-J, Di Primo C: Aptamer selection by direct microfluidic recovery and surface plasmon resonance evaluation, *Biosens Bioelectron* 80:418–425, 2016.

Dostálek J, C̆tyroký J, Homola J, Brynda E, Skalský M, Nekvindová P, Špirková J, Škvor J, Schröfel J: Surface plasmon resonance biosensor based on integrated optical waveguide, *Sens Actuators B Chem* 76:8–12, 2001.

Dutra RF, Kubota LT: An SPR immunosensor for human cardiac troponin T using specific binding avidin to biotin at carboxymethyldextran-modified gold chip, *Clin Chim Acta* 376:114–120, 2007.

Fang S, Lee HJ, Wark AW, Corn RM: Attomole microarray detection of microRNAs by nanoparticle-amplified SPR imaging measurements of surface polyadenylation reactions, *J Am Chem Soc* 128:14044–14046, 2006.

Fasoli JB, Corn RM: Surface enzyme chemistries for ultrasensitive microarray biosensing with SPR imaging, *Langmuir* 31:9527–9536, 2015.

Granqvist N, Liang H, Laurila T, Sadowski J, Yliperttula M, Viitala T: Characterizing ultrathin and thick organic layers by surface plasmon resonance three-wavelength and waveguide mode analysis, *Langmuir* 29:8561–8571, 2013.

Gwon HR, Lee SH: Spectral and angular responses of surface plasmon resonance based on the Kretschmann prism configuration, *Mater Trans* 51:1150–1155, 2010.

Homola J: Present and future of surface plasmon resonance biosensors, *Anal Bioanal Chem* 377:528–539, 2003.

Homola J: Surface plasmon resonance sensors for detection of chemical and biological species, *Chem Rev* 108:462–493, 2008.

Homola J, Yee SS, Gauglitz G: Surface plasmon resonance sensors: review, *Sens Actuators B Chem* 54:3–15, 1999.

Homola J, Dostálek J, Chen S, Rasooly A, Jiang S, Yee SS: Spectral surface plasmon resonance biosensor for detection of staphylococcal enterotoxin B in milk, *Int J Food Microbiol* 75:61–69, 2002.

Kamimori H, Hall K, Craik DJ, Aguilar M-I: Studies on the membrane interactions of the cyclotides kalata B1 and kalata B6 on model membrane systems by surface plasmon resonance, *Anal Biochem* 337:149–153, 2005.

Karlsson R: SPR for molecular interaction analysis: a review of emerging application areas, *J Mol Recognit* 17:151–161, 2004.

Kretschmann E, Raether H: Radiative decay of non radiative surface plasmons excited by light, *Z Naturforsch* 23a:2135–2136, 1968.

Lawrence CR, Geddes NJ, Furlong DN, Sambles JR: Surface plasmon resonance studies of immunoreactions utilizing disposable diffraction gratings, *Biosens Bioelectron* 11:389–400, 1996.

Lee J-H, Huh Y-M, Jun Y, Seo J, Jang J, Song H-T, Kim S, Cho E-J, Yoon H-G, Suh J-S, Cheon J: Artificially engineered magnetic nanoparticles for ultra-sensitive molecular imaging, *Nat Med* 13:95–99, 2006.

Liang H, Miranto H, Granqvist N, Sadowski JW, Viitala T, Wang B, Yliperttula M: Surface plasmon resonance instrument as a refractometer for liquids and ultrathin films, *Sens Actuators B Chem* 149:212–220, 2010.

Liu Y, Liu Q, Chen S, Cheng F, Wang H, Peng W: Surface plasmon resonance biosensor based on smart phone platforms, *Sci Rep* 5:12864, 2015.

Löfås S, Malmqvist M, Rönnberg I, Stenberg E, Liedberg B, Lundström I: Bioanalysis with surface plasmon resonance, *Sens Actuators B Chem* 5:79–84, 1991.

Lopez F, Pichereaux C, Burlet-Schiltz O, Pradayrol L, Monsarrat B, Estève J-P: Improved sensitivity of biomolecular interaction analysis mass spectrometry for the identification of interacting molecules, *Proteomics* 3:402–412, 2003.

McDonnell JM: Surface plasmon resonance: towards an understanding of the mechanisms of biological molecular recognition, *Curr Opin Chem Biol* 5:572–577, 2001.

Miyazaki CM, Pereira TP, Mascagni DBT, de Moraes ML, Ferreira M: Monoamine oxidase B layer-by-layer film fabrication and characterization toward dopamine detection, *Mater Sci Eng C* 58:310–315, 2016.

Morgan H, Taylor DM: A surface plasmon resonance immunosensor based on the streptavidin–biotin complex, *Biosens Bioelectron* 7:405–410, 1992.

Nylander C, Liedberg B, Lind T: Gas detection by means of surface plasmon resonance, *Sens Act* 3:79–88, 1982–1983.

Oh B-K, Kim Y-K, Park KW, Lee WH, Choi J-W: Surface plasmon resonance immunosensor for the detection of *Salmonella typhimurium, Biosens Bioelectron* 19:1497–1504, 2004a.

Oh B-K, Lee W, Kim Y-K, Lee WH, Choi J-W: Surface plasmon resonance immunosensor using self-assembled protein G for the detection of *Salmonella paratyphi, J Biotechnol* 111:1–8, 2004b.

Olaru A, Bala C, Jaffrezic-Renault N, Aboul-Enein HY: Surface plasmon resonance (SPR) biosensors in pharmaceutical analysis, *Crit Rev Anal Chem* 45:97–105, 2015.

Otto A: Excitation of nonradiative surface plasma waves in silver by the method of frustrated total reflection, *Z Phys* 216:398–410, 1968.

Pei R, Cui X, Yang X, Wang E: Assembly of alternating polycation and DNA multilayer films by electrostatic layer-by-layer adsorption, *Biomacromolecules* 2:463–468, 2001.

Pieper-Fürst U, Stöcklein WFM, Warsinke A: Gold nanoparticle-enhanced surface plasmon resonance measurement with a highly sensitive quantification for human tissue inhibitor of metalloproteinases-2, *Anal Chim Acta* 550:69–76, 2005.

Piliarik M, Homola J: Surface plasmon resonance (SPR) sensors: approaching their limits? *Opt Express* 17:16505–16517, 2009.

Ritchie RH, Arakawa ET, Cowan JJ, Hamm RN: Surface-plasmon resonance effect in grating diffraction, *Phys Rev Lett* 21:1530–1533, 1968.

Scarano S, Mascini M, Turner APF, Minunni M: Surface plasmon resonance imaging for affinity-based biosensors, *Biosens Bioelectron* 25:957–966, 2010.

Singh VV, Gupta G, Batra A, Nigam AK, Boopathi M, Gutch PK, Tripathi BK, Srivastava A, Samuel M, Agarwal GS, Singh B, Vijayaraghavan R: Greener electrochemical synthesis of high quality graphene nanosheets directly from pencil and its SPR sensing application, *Adv Funct Mater* 22:2352–2362, 2012.

Singh M, Holzinger M, Tabrizian M, Winters S, Berner NC, Cosnier S, Duesberg GS: Noncovalently functionalized monolayer graphene for sensitivity enhancement of surface plasmon resonance immunosensors, *J Am Chem Soc* 137:2800–2803, 2015.

Situ C, Mooney MH, Elliott CT, Buijs J: Advances in surface plasmon resonance biosensor technology towards high-throughput, food-safety analysis, *Trends Anal Chem* 29:1305–1315, 2010.

Vashist SK, Dixit CK, MacCraith BD, O'Kennedy R: Effect of antibody immobilization strategies on the analytical performance of a surface plasmon resonance-based immunoassay, *Analyst* 136:4431, 2011.

Wagner T, Roth S: Surface-plasmon-resonance investigations of Langmuir–Blodgett films of donor–acceptor substituted polyenes: linear optical and electro-optic properties, *Synth Met* 54:307–314, 1993.

Wang S, Huang X, Shan X, Foley KJ, Tao N: Electrochemical surface plasmon resonance: basic formalism and experimental validation, *Anal Chem* 82:935–941, 2010.

Zalewska M, Kochman A, Estève J-P, Lopez F, Chaoui K, Susini C, O yhar A, Kochman M: Juvenile hormone binding protein traffic—interaction with ATP synthase and lipid transfer proteins, *Biochim Biophys Acta* 1788:1695–1705, 2009.

Index